D0736310

Mammalogy
Techniques Manual

James M. Ryan

LIS - LIBRARY

Date	Fund
15/7/14	b1-Che

Order No.

2485540

University of Chester

WITHDRAWN

2nd Edition

Copyright © 2011 by James M. Ryan

Cover Photo by Julie Lubik/Shutterstock.com

Book design by James M. Ryan

All rights reserved.

Published by: Lulu, 3101 Hillsborough St., Raleigh, NC, 27607

ISBN 978-1-257-83194-4

No part of this book may be reproduced in any form or by any electronic or mechanical means including information storage and retrieval systems, without permission in writing from the author. The only exception is by a reviewer, who may quote short excerpts in a review.

First edition: December 2010

Second edition: July 2011

Please cite as: Ryan, J. M. 2011. Mammalogy Techniques Manual.

2nd edition, Lulu, Raleigh, NC

KT-468-617

Preface

The purpose of this manual is to provide students with an opportunity to practice some of the techniques used by today's mammalogists. An additional goal is to get students outdoors, where they can hone their observation skills and develop the essential skills to become the next generation of practicing mammalogists.

Mammalogists are keenly aware of the need to get student back into the field. In 2007 mammalogist Mark Hafner published a paper lamenting the decline in field-based studies of natural history and mammalogy over the last few decades. He noted that "We can no longer assume that our students, many of whom are nature-deprived, have a set of basic outdoor skills that will safeguard them in the field; these students need to be taught about fieldwork from the ground up" (Mark S. Hafner 2007 Field Research in Mammalogy: An Enterprise in Peril. *Journal of Mammalogy*: 88, 1119-1128.). This manual is meant to address some these concerns by providing a set of modern field and laboratory techniques that every student of mammalogy should know.

Obviously, there is much that is not covered. For example, there are no labs on mammal identification. This is because most textbooks cover those details and because it would be impossible to include such information for all regions in North America (or the World). The assumption is that instructors who wish to include identification of regional mammals will do so on their own.

Where software is introduced to ease analysis of large or complex data sets, Open Source (free) or Web-based tools are the preferred choice. The primary reason is to minimize cost to students. In addition, faculty at urban institutions with little access to field sites should be able to introduce many of these labs without collecting data from the field using data sources provided with the exercises, or available at:

http://www.wildmammal.com/downloads

Unless otherwise indicated, all drawings and photographs were produced by the author.

Acknowledgements

I am grateful to the following individuals who provided data or editorial suggestions for this manual. Dr. Rick Mace generously provided the raw GPS data for tracking a female grizzly bear in Chapter 11. A subset of that data is used in the exercises in this manual. Dr. Roland Kays provided valuable suggestions and editorial comments on several chapters in this manual. I am very grateful to Dr. Larry Phelps from the University of Wisconsin Baraboo (retired), who generously provided the human karyotype photos. I also offer my sincerest gratitude to my wife Gillian for her patience and support.

Chapters

Using CAPTURE and JOLLY Software for Mark-Recapture Data

Transects - Distance Sampling Using DISTANCE 121

1

Mammal Skulls

TIME REQUIRED

1 lab period of 3-4 hours

LEVEL OF DIFFICULTY

Basic

LEARNING OBJECTIVES

Learn the anatomy of the mammalian skull.

Use the anatomical terms in a taxonomic key to identify mammalian skulls.

Understand the structure of a dichotomous key.

Use the traits in the skull key to identify unknown specimens.

EQUIPMENT REQUIRED

A collection of mammal skulls from a variety of mammalian Orders.

A collection of bolts, nuts, and screws.

Callipers or other measurement tools.

BACKGROUND

There are over 4,000 living species of mammals grouped into 29 Orders. Because many species (especially species of small mammals) appear similar to the untrained eye, scientists have developed dichotomous keys to facilitate identification of species. Dichotomous keys are pairs of statements that allow the user to easily identify an unknown organism. The key is arranged in "couplets" consisting of two statements, which ask the user to make a choice about the presence, absence, or aspect of a particular characteristic of their unknown organism. That choice leads the user to a new "couplet" in the key and the process repeats until a "couplet" choice leads to the name of the organism (or the group to which it belongs). Each "couplet" ends in either a number indicating which "couplet" to go to next in the key, or in the name of the taxa. It is good practice to record your path through the key, including the final identification of your taxa (e.g. 1a; 2b; 3b; 4a; African elephant), so that you may retrace your steps if you make a mistake or if the key is long and complex.

Most dichotomous keys use scientific terminology to describe features. Thus, it is often important to familiarize yourself with basic anatomical terminology before attempting to use a key. Mammal skulls are often used in keys to identify species. However, mammal skulls are complex structures comprised of numerous bones, processes, holes and teeth (usually). Each is important and has been shaped by natural selection. For example, the diverse feeding habits of mammals are reflected in their teeth and jaw morphology. Teeth may be specialized for grasping, crushing, or grinding; the position and structure of the teeth often provide valuable clues to the animal's diet (see Chapter 2).

BONES AND FEATURES OF THE SKULL

The major bones of the mammalian skull are listed below and illustrated in Figures 1.1, 1.2, and 1.3.

Alisphenoid	Pterygoid
Orbitosphenoid	Lacrimal
Dentary (bone of the mandible)	Basisphenoid
Palatine	Squamosal (Temporal)
Ethmoid	Maxilla
Parietal	Tympanic bullae (composite)
Frontal	Nasal
Premaxilla	Vomer
Interparietal	Occipital
Presphenoid	Zygomatic arch (composite)
Jugal	

BONE FEATURES

The following list of terms provide a basic introduction; your instructor may supplement this list with additional terms.

Crest, a narrow prominent ridge.

Condyle, a smooth rounded projection for articulation with another bone.

Foramen, a hole in a bone (typically, passes a nerve or blood vessels).

Fossa, a shallow depression or trench on the surface of a bone.

Labial, on the side adjacent to the lips.

Lingual, on the side adjacent to the tongue.

Line, a narrow raised ridge.

Meatus, a small tubular opening.

Process, a small projection or bump.

Septum, a bony fence that separates two regions.

Sulcus, a groove.

Suture, the line formed by the junction of two bones.

Symphysis, the cartilaginous junction or articulation formed between two bones; this junction or articulation may fuse (e.g. the two dentary bones fuse to form the mandibles at the mandibular symphysis).

Trochanter, a large rounded projection for muscle attachment.

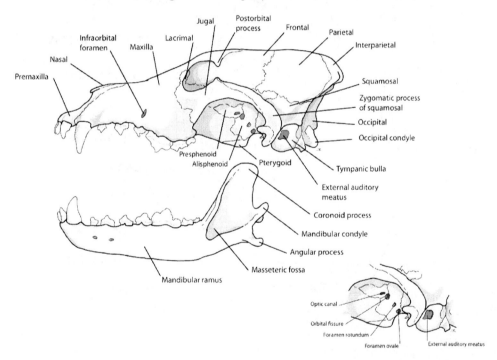

Figure 1.1 Lateral view of a dog skull illustrating the main bones and features. The enlarged section shows the position of foramina in the orbital region.

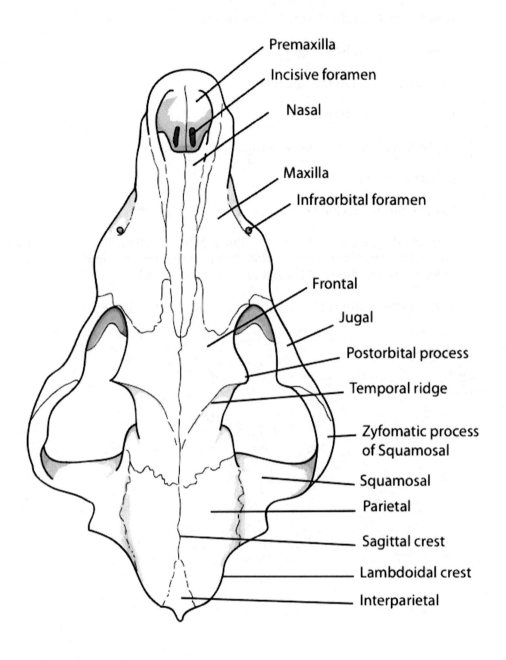

Premaxilla
Incisive foramen
Nasal
Maxilla
Infraorbital foramen
Frontal
Jugal
Postorbital process
Temporal ridge
Zyfomatic process of Squamosal
Squamosal
Parietal
Sagittal crest
Lambdoidal crest
Interparietal

Figure 1.2 Dorsal view of a dog skull illustrating the main bones and features.

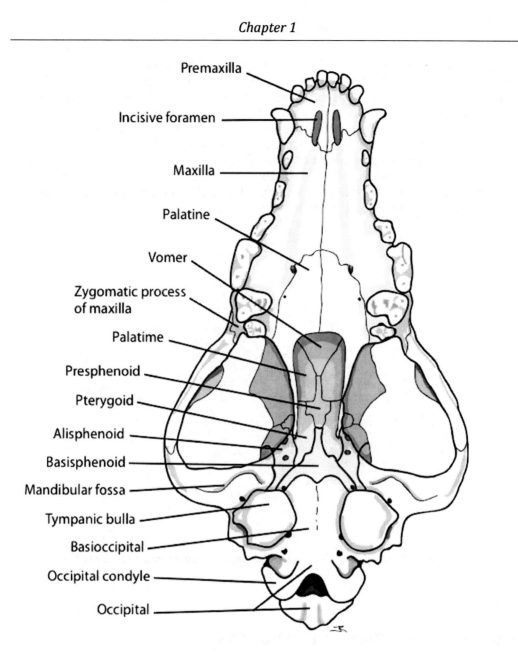

Premaxilla

Incisive foramen

Maxilla

Palatine

Vomer

Zygomatic process
of maxilla

Palatime

Presphenoid

Pterygoid

Alisphenoid

Basisphenoid

Mandibular fossa

Tympanic bulla

Basioccipital

Occipital condyle

Occipital

Figure 1.3 Ventral view of a dog skull illustrating the main bones and features.

VARIATION IN MAMMALIAN SKULLS

Mammal skulls vary tremendously in size and shape. Therefore, it is important to recognize the position of each bone regardless of species. Fortunately, the bones retain the same relative positions. For example, compare the canid skulls in Figures 1.1 through 1.3 with those of the pronghorn (*Antilocapra americana*) in Figures 1.4 to 1.6. Notice that the nasal bones still form the anterior margin of the nasal opening and are bordered by the premaxilla and maxilla (laterally) and the frontals (posteriorly). Likewise, the palatine bone is still posterior to the maxilla on the roof of the mouth (palate). Also notice that the pronghorn has unusually large orbits, projecting away from the skull. Their large eyes give them a very wide field of view for detecting predators. Artiodactyls tend to have long preorbital skulls (rostrums) formed by elongated nasals and frontals. The frontal bones typically bear horn cores (sometimes with the horn sheath still attached) or antler peduncles (in adult males of the family Cervidae). A postorbital bar is usually also present, serving to fully enclose the orbit. Another feature common to many artiodactyls is the loss of upper incisors and canines (retained in some artiodactyls).

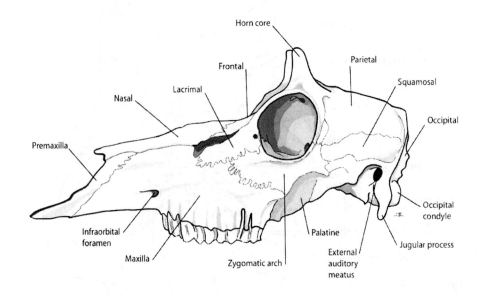

*Figure 1.4 Lateral view of the skull of a pronghorn (*Antilocapra americana*).*

ZYGOMATIC MORPHOLOGY IN RODENTS

Rodents also have a great deal of variation in their skull morphology. A typical rodents skull (*Peromyscus*) is shown in Figure 1.7. Rodents (Order Rodentia) are characterised by a pair of continuously growing upper and lower incisors. The incisors are used for gnawing, and are bevelled to a chisel-shape at the tips. The constant wear on the incisors is necessary because these teeth are evergrowing. If, for example, one incisor in an upper pair is broken, and no longer able to occlude with the lower incisor, then the opposing lower incisor will continue to grow. In some circumstances this incisor grows so far as to prevent the mouth from closing and the animal starves to death. Rodent incisors are separated from the cheekteeth by a long diastema. The nasals are long and narrow, and in many species extend anterior to the incisors.

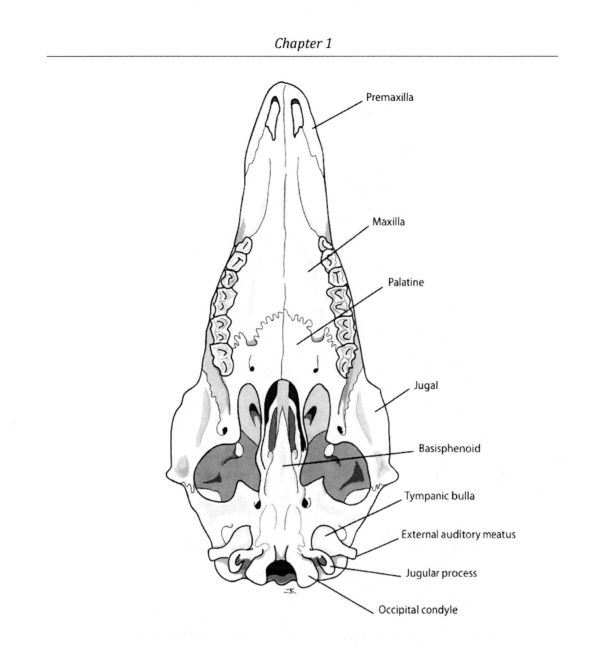

Premaxilla

Maxilla

Palatine

Jugal

Basisphenoid

Tympanic bulla

External auditory meatus

Jugular process

Occipital condyle

Figure 1.5 Ventral view of the skull of a pronghorn (Artiodactyla, Antilocapra americana*).*

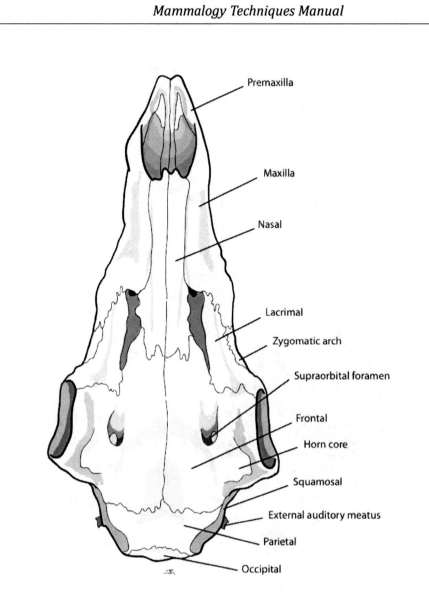

Figure 1.6 Dorsal view of the skull of a pronghorn (Artiodactyla, Antilocapra americana).

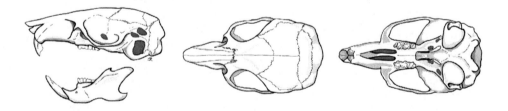

Figure 1.7 Lateral, dorsal, and ventral views of the skull of a Peromyscus (Rodentia) showing the bevelled incisors and diastema.

The upper and lower incisors come together during gnawing; their sharp, bevelled edges allow the incisors to chisel away hard materials. Such gnawing requires powerful jaw muscles that act to move the lower jaw forward, bringing the lower incisors into occlusion with the uppers. Rodents have evolved unique arrangements of jaw muscles to accomplish this task. Primitively, the masseter muscle was divided into superficial, lateral, and medial masseters, with the superficial masseter originating near the anterior end of the upper toothrow, the lateral masseter originating from the middle of the zygomatic arch, and the smaller medial masseter originating along the inside of the zygomatic arch (Figure 1.8). In this primitive configuration, the incisors could be used for only relatively simple gnawing movements. Rodents with this ancestral condition are called **protrogomorphous**. Protrogomorphy was common among early rodents, but only the mountain beaver (or sewell), *Aplodontia rufa*, retains this zygomassetric arrangement among living rodents.

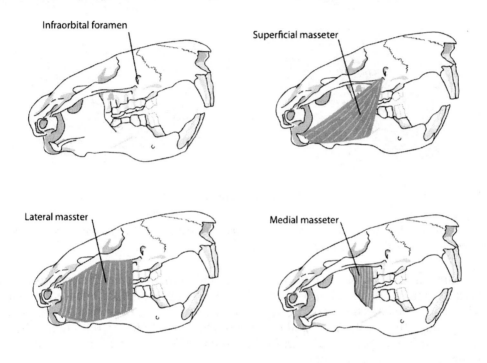

Protrogomorphous masseter

Figure 1.8 The masseter arrangement in protrogomorphous (ancestral) rodents, showing the relative size of the infraorbital foramen and the positions of the masseter slips.

As rodents continued to rely on gnawing, selection favoured a more anterior position for at least some slips of the masseter. By moving a part of the masseter onto the rostrum in front of the zygomatic arch, rodents gained mechanical advantage during gnawing. These changes set the stage for the rapid radiation of rodents into the many species we recognize today. However, rodents accomplished this anterior repositioning of the jaw muscles in at least three different ways.

Sciuromorphous rodents (Figure 1.9) have the lateral masseter originating from a broad region at the front of the zygomatic arch (termed the zygomatic plate). The medial masseter is larger and extends posteriorly along the zygomatic arch (relative to that in protrogomorphous masseters). This condition is found in beavers, squirrels, gophers, kangaroo rats and their allies (Family Heteromyidae), and several other rodent families.

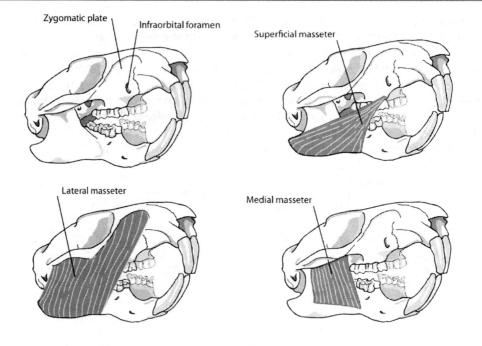

Sciuromorphous masseter

Figure 1.9 The masseter arrangement in sciuromorphous rodents, showing the relative size of the infraorbital foramen and the positions of the masseter slips.

The infraorbital foramen is greatly enlarged in the **hystricomorphous** masseter arrangement (Figure 1.10). The greatly expanded portion of the medial masseter originates on the lateral wall of the upper rostrum and passes ventrally and posteriorly through the greatly enlarged infraorbital foramen to insert on the mandible. Hystricomorphous masseters occur in New and Old World porcupines, guinea pigs, and many other rodent families.

A third **myomorphous** condition also involves a slip of the masseter passing through an enlarged infraorbital foramen (Figure 1.10). In this case, however, there is a zygomatic plate below an enlarged (modestly) infraorbital foramen. It is worth noting that mammalogists still aren't sure how each masseter arrangement evolved, but it is clear that rodents independently evolved several solutions to the problem of repositioning the masseters for gnawing.

Today, mammalogists rely less on masseter morphology for classifying rodents and more on the morphology of the angular process of the lower jaw (e.g. hystricognathy versus sciurognathy) and on other morphological and molecular characters. In the hystricognathous condition, the angular process of the dentary is lateral to the plane of the tooth row, whereas in the sciurognathous jaw the angular process is approximately in a line with the rest of the toothrow (Figure 1.11).

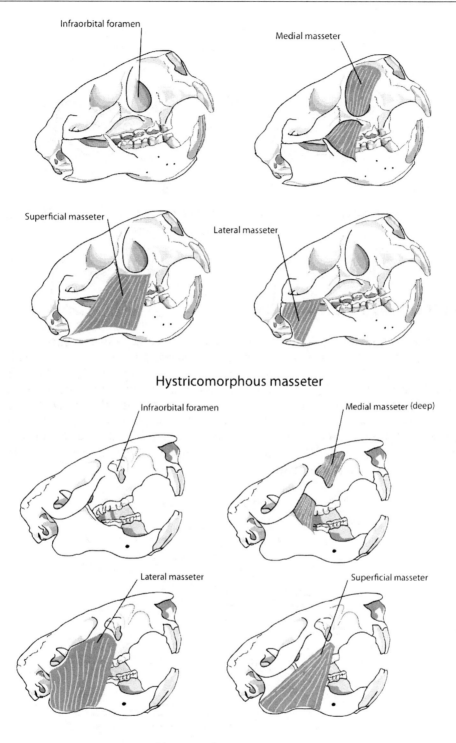

Hystricomorphous masseter

Myomorphous masseter

Figure 1.10 The masseter arrangement in hystricomorphous (top) and myomorphous (bottom) rodents.

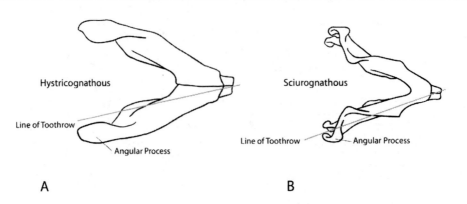

Figure 1.11 A comparison of the position of the angular process of the lower jaw in (A) a hystricognathous rodent and (B) a sciurognathous rodent.

TELESCOPING IN CETACEANS

Another unusual skull morphology evolved in cetaceans in response to a fully aquatic lifestyle. Cetaceans may have highly telescoped and/or asymmetrical skulls. Toothed whales (Odontocetes) have telescoped skulls (Figure 1.12), where the external nares are repositioned on the top of the skull. This allows these cetaceans to breathe with just the nostrils (blow holes) exposed above the water. Telescoping involves the elongation of the rostrum and the posterior displacement of the nasals, maxillae, and premaxillae (among others) relative to the braincase. Odontocetes also have asymmetrical skulls. The bones on the right side of the skull are larger than their counterparts on the left side.

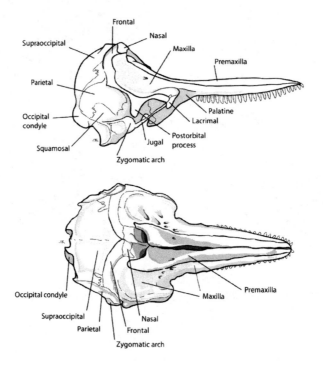

Figure 1.12 The asymmetrical and telescoped skull of a bottlenosed dolphin (Tursiops truncatus).

SKULL MEASUREMENTS

Measuring skulls can aid in species identification, especially where skulls are nearly identical in shape but vary in size (Martin, et al., 2000). Mammalogists use a variety of standard skull measurements to determine sex, establish age or maturity, and for geographic comparisons (Figures 1.13 to 1.15). Measurements are typically taken with dial callipers (for increased accuracy). Important standard measurements are:

Greatest length of skull (greatest skull length, total skull length, maximum length of skull)= from the most anterior part of the rostrum (excluding teeth) to the most posterior point of the skull.

Greatest depth of skull = greatest length from the ventral most margin of the skull (including teeth) to the dorsal most margin of the skull.

Occipitonasal length = distance from the anterior most tip of nasal bone to the posterior most margin of the occipital bone.

Length of maxillary toothrow = distance from the anterior margin of the upper canine tooth to the posterior most margin of the last molar on the maxilla.

Length of mandibular toothrow = distance from the anterior margin of the lower canine tooth to the posterior most margin of the last molar on the mandible.

Mandible length = total length of the mandible from the alveolus of the incisors to the posterior most margin of the mandible.

Basal length = from the anterior most margin of the premaxilla to the anterior most margin of the foramen magnum.

Breadth of braincase = greatest width across the braincase posterior to the zygomatic arches.

Condylobasal length = from a line connecting the posterior most projections of the occipital condyles to the anterior edge of the premaxillae.

Incisive foramina length = greatest length of the anterior palatal (incisive) foramina.

Palatal length = from the anterior edge of premaxilla to the anterior most point on the posterior edge of palate.

Postpalatal length = From the anterior most margin of the posterior edge of palate to the anterior most margin of the foramen magnum.

Zygomatic breadth = greatest distance between the outer margins of the zygomatic arches.

Interorbital breadth = least distance across the top of the skull anterior to the postorbital process/bar.

Postorbital breadth = greatest distance across the top of the skull at the level of the postorbital process/bar.

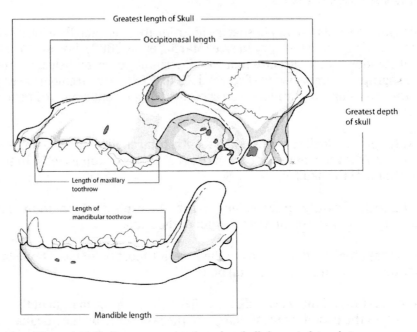

Figure 1.13 *Skull measurements for a dog skull shown in lateral view.*

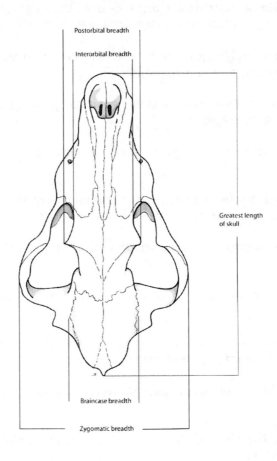

Figure 1.14 *Skull measurements for a dog skull shown in dorsal view.*

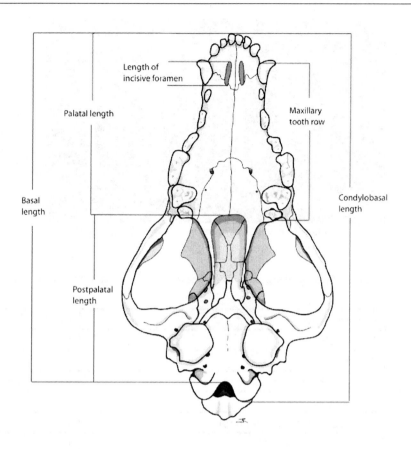

Figure 1.15 Skull measurements for a dog skull shown in ventral view.

Exercises

EXERCISE 1: THE NUTS & BOLTS

As an example of how to construct and use a dichotomous key, you will begin with a set of simple "organisms" like those shown in Figure 1.16. These "organisms" are various sizes and shapes of nails, screws, and bolts (those provided by you instructor may differ).

To begin organizing this set of "organisms" into a key, first divide the set into two groups based on some easily identifiable feature. Record the two choices in the format of a dichotomous key and repeat the process until all of the "organisms" have been identified (Table 1.1).

Your instructor will provide you with a set of hypothetical "organisms" similar to those in the key above. Working individually or with a partner create a dichotomous key based on the characteristics of various organisms. Use easy to identify and describe characters in your key. When you are finished, your key will be given to another group and they will use it to attempt to identify the "organisms."

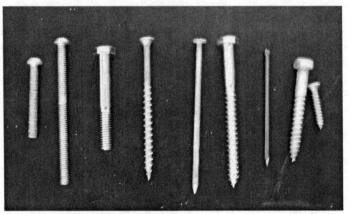

Figure 1.16 A set of "organisms" used in the example key.

Table 1.1 A sample dichotomous key for the "organisms" shown in Figure 1.16.

	Character	Go to
1A	Straight shaft	2
1B	Threaded shaft	3
2A	Flattened head at top of shaft	Common nail
2B	Head not flattened	Finishing nail
3A	Head dome-shaped (rounded)	4
3B	Head not dome-shaped	5
4A	Shaft more than 10 time the width of head	Long bolt
4B	Shaft less than 10 times the width of head	Short bolt
5A	Head cone-shaped (not hexagonal)	6
5B	Head hexagonal (six-sided)	7
6A	Shaft entirely threaded	Wood screw
6B	Shaft not threaded near head	Deck screw
7A	Shaft pointed at tip	8
7B	Shaft blunt at tip	Carriage bolt
8A	Shaft short and head thick (relatively)	Small lag screw
8B	Shaft long and head thin (relatively)	Long lag screw

STEPS

1. Obtain a group "organisms".

2. Assign each "organism" a unique name or letter.

3. Carefully examine or think about the characteristics of each organism.

4. Divide the group of "organisms" into two piles based on the presence or absence of a trait and construct the first couplet in the key based on that trait. Use only one page for your key.

5. Make a second page as an BLANK key. In each spot on the right side where you have the final identity of an "organism," you should leave a blank for the other groups to fill in (e.g. do not list the name of the organism).

6. When you are finished, you will be given the key created by another group. Try to identify each "organism" and fill in the blanks in their key by following the key precisely.

7. Assign the other groups key a score: A score of 1 is very poor and a score of 5 is excellent. Also provide an explanation for your score, by answering the following questions:

- Where were there difficulties in the way traits were described in the key?

- How might those difficult areas be reworded to improve the meaning?

- If you were given a new "organism" that you had not used in the key (but similar), could the key be easily modified to add the new "species"?

EXERCISE 2: DICHOTOMOUS KEYS OF SKULLS

Now you will put theory into practice. You will be given a series of mammal skulls by your instructor. You should look them over carefully. Working individually or with a partner create a dichotomous key of these skulls by using the skull features and measurements described in this lab manual. Use easy to identify and describe characters in your key, but you must use appropriate terminology or measurement descriptions. For example, if you want to divide them into two groups of skulls based on size, you would measure the greatest length of skull for each skull and make a determination as to where to cut the length to define the two groups. You might say:

2A Greatest length of skull is greater than 4 cm.

2B Greatest length of skull is less than 4 cm.

Steps in constructing the key

1. Obtain a group skulls.

2. Assign each skull a unique letter.

3. Carefully examine or think about the characteristics of each skull.

4. Divide the group of skulls into two piles based on the presence or absence of a trait and construct the first couplet in the key based on that trait.

5. Repeat the process until all of the skull have been entered into the key.

6. Use the blank worksheets provided.

EXERCISE 3: MYSTERY MAMMAL SKULL

Your instructor will provide you with several "mystery" skulls for you to identify. You will also be given a copy of *A Key to the Skulls of North American Mammals* by Bryann Glass and Monte Thies (1997) to help you in keying out your mystery skulls. The key is divided into two parts; the first part is a key to mammalian Orders (pages 12 -13). Once you know the Order, you will turn to the specific key to that Order. For example if you have keyed out a skull as belonging to the Order Carnivora, then you would turn to the key beginning on page 29 and continue. (Note: the Order Insectivora is no longer used in modern mammalian classifications, see Vaughan, Ryan, and Czaplewski 2010 for details). Be sure to keep a record of the path you follow through the key. This is important; without a path, it is difficult to retrace your steps when resolving misidentifications. Use the glossary and figures provided in Glass and Thies, 1997 and this chapter. If terms or descriptions remain confusing, ask your instructor.

Bibliography

Glass, B. P. and M. L. Thies. (1997) *A Key to the Skulls of North American Mammals.* 3rd edition. (Self-published)

Martin, R., Pine, R., and A. DeBlase. (2000) *A Manual of Mammalogy.* 3rd edition, McGraw-Hill. Columbus, OH.

Vaughan, T., Ryan, J.M., and N. Czaplewski. (2010) *Mammalogy*, 5th edition. Jones & Bartlett Publishing, Sudbury, MA.

Appendix

Skull ID	
Path Followed	
Name of Mammal	
Skull ID	
Path Followed	
Name of Mammal	
Skull ID	
Path Followed	
Name of Mammal	

Couplet	Description	Go to

2

Mammalian Teeth

TIME REQUIRED

One lab period of 3-4 hours

LEVEL OF DIFFICULTY

Basic

LEARNING OBJECTIVES

To understand the basic evolution of mammalian teeth.

Recognize cusp patterns on mammalian teeth.

Link diet to tooth morphology in mammalian species.

Identify the different types of teeth: incisors, canines, premolars & molars.

Understand the diversity in mammalian cheek teeth.

Write dental formulas for mammalian species.

EQUIPMENT REQUIRED

A collection of mammal skulls from a variety of Orders

BACKGROUND

Teeth are extremely important to mammals (and to mammalogists). Modifications in the dentition and masticatory apparatus appear early in the diversification of mammals (from their therapsid ancestors) (Figure 2.1). These alterations reflect changes in physiology and behavior of ancestral mammals. Early mammals were small and probably foraged during the cooler hours from evening to dawn. Their small body size and nocturnal habits favored a consistently higher body temperature, which in turn required the ability to capture, chew, and digest food efficiently. The lineage leading to modern mammals is characterized by three important trends in dentition:

1) reduction in the total number of teeth.

2) reduction in the number of generations of teeth (e.g. from polyphyodonty to diphyodonty or monophyodonty).

3) increased morphological complexity of teeth within the toothrow (e.g. heterodonty).

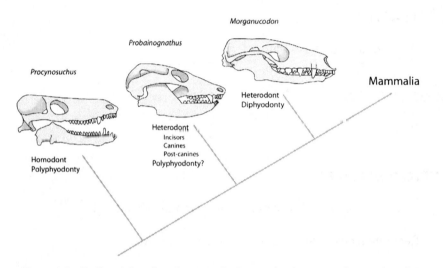

Figure 2.1. Skull and dental evolution in the lineage leading to modern mammals.

As the Mammalia continued to diversify, some of these trends were reversed; some odontocete whales and dolphins have more than 200 teeth and those teeth are secondarily homodont. A few mammal species have lost teeth altogether. For most mammals, however, teeth have become variously specialized for snipping, grinding, crushing, puncturing, or slicing food. These diverse ways of processing food are reflected in the morphology of the teeth (see Vaughan et al., 2010). For example, the molars of herbivores are flattened and the occlusal surface is used to grind plant material. In contrast, the molars of carnivores (e.g. the hypercarnivorous felids) are blade-like and used to slice meat into more manageable pieces prior to swallowing. Indeed, we can often deduce the diet of the animal from its teeth alone. Finally, much of what we know about the evolutionary history of mammals we have learned from teeth. This is because teeth are exceedingly hard and fossilize better than any other body part. Indeed, many species of fossil mammals are known only from their teeth. Fortunately, although teeth vary widely between species, they show very little variation within species. This feature makes them useful in reconstructing the phylogenetic history of mammals.

INTERNAL STRUCTURE

Regardless of their shape, teeth share a common internal structure. The surface, or crown, is covered with a very hard substance called enamel. Just below the enamel is a layer of dentine. Dentine is hard, but not as hard as enamel. This differential hardness may result in surface ridges on the tooth, where the enamel and dentine are both exposed at different heights on the tooth's surface (Figure 2.2). The layer of dentine surrounds the pulp cavity. Pulp is a cellular tissue containing odontoblast cells, blood vessels and nerves. The vessels and nerves enter the tooth via a canal at the base of the tooth (e.g. the root). The odontoblast cells in the pulp secrete dentine. The tooth is anchored in a socket in the jaw with a material called cementum. Despite the name, cementum does not act as a glue (sitting on the surface), but instead contains tiny connective tissue fibers rooted to the bone. Cementum may also extend onto the crown of the tooth in some species.

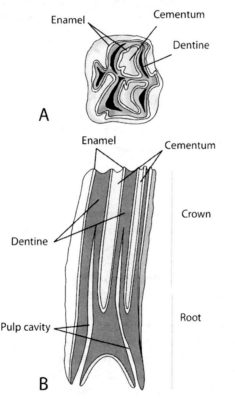

Figure 2.2 (A) occlusal view of the surface of a horse molar showing the relationship of the enamel ridges and the depressions formed by the softer dentine. (B) sagittal section through a horse molar, showing the internal structures of the tooth.

In most mammals the root contains a canal through which the blood vessels and nerves enter the pulp cavity from below. When the root canal is open the tooth continues to grow, but when the animal matures, the root canal typically closes and the tooth stops growing. Rodent incisors are one example of teeth that continue to grow throughout the rodents life (i.e. root-less or ever-growing teeth).

KINDS OF TEETH

Teeth are present on three bones in mammals: the maxilla and premaxilla of the upper jaw, and the dentary of the lower jaw (Figure 2.3).

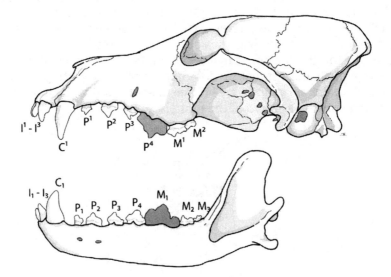

Figure 2.3   A dog skull illustrating a typical complement of teeth. The shaded upper premolar 4 and the lower molar 1 represent the carnassial pair in many carnivores.

The ancestral homodont dentition has given way to four distinct kinds of teeth in many mammals: incisors, canines, premolars, and molars.

Incisors are the most anterior teeth, located in the premaxilla of the upper jaw and in the dentary of the lower jaw. Incisors are generally simple teeth, used for grasping and clipping. They are modified in many mammals. Lemurs have modified incisors into a "toothcomb" used for grooming. Rodents have chisel-shaped, evergrowing incisors used for gnawing. Perhaps the most spectacularly modified incisors are the tusks of elephants.

Canines are posterior to the incisors. Typically a single canine in present in each quadrant and they are the first tooth in the maxilla. Canines are simple, conical teeth with a single cusp and a single root. They are used for stabbing or piercing. Rodents, many artiodactyls, and a few other taxa lack canines or have canines only in the lower jaw. Narwhals, musk deer, and the extinct saber-toothed cats have enlarged canines that probably serve other functions beyond feeding (e.g. social signalling, defensive weapons).

Premolars lie immediately posterior to the canines in the maxilla and dentary. Premolars are deciduous, whereas molars are not replaced. Premolars vary from tiny pegs (shrews) to massive crushing teeth of many herbivores (Figures 2.4 & 2.5).

Molars are the most posterior teeth in the jaws. They also vary tremendously in size and shape, but exist only as adult teeth.

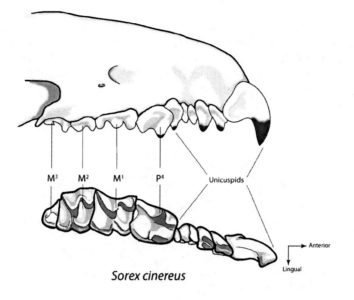

Sorex cinereus

Figure 2.4   Rostrum of a shrew, Sorex cinereus. The incisors, canines, and anterior premolars are called unicuspid teeth.

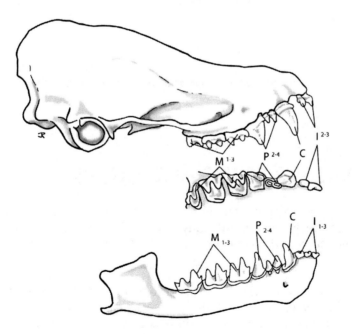

Myotis lucifugus

Figure 2.5   Skull and lower jaw of a little brown bat, Myotis lucifugus.

Mammalogists use a number of terms to refer to the orientation of tooth surfaces. The occlusal surface is the exposed surface that meets with the tooth in the opposite toothrow. The labial side of the tooth is adjacent to the lips, and the lingual side is adjacent to the tongue. Anterior and posterior have their traditional meaning when referring to teeth.

While mammalian ancestors evolved a complex series of teeth, each suited to a specialized function, some mammals have secondarily evolved homodont teeth. Homodont teeth are all similar is shape. Homodont teeth in modern mammals tend to be found in fully aquatic piscivores (fish-eaters) such as toothed-whales (Figure 2.6) and several species of pinnipeds.

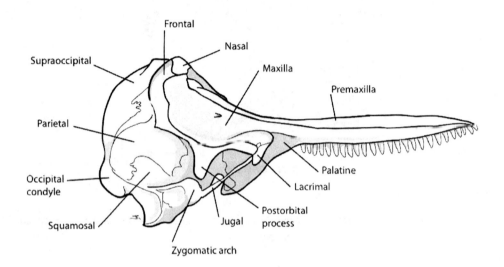

Figure 2.6   Lateral view of the homodont dentition of a dolphin, Tursiops truncatus.

OCCLUSAL PATTERNS AND CUSPS

The occlusal surface of an erupted tooth has patterns of cusps and lophs that are species specific and often highly informative. Cusp morphology and position are correlated with diet and are the result of evolutionary and developmental constraints.

Early mammals had teeth with three major cusps in a relatively straight line (see Vaughan et al., 2010 for details). These cusps increased the slicing surfaces and provided an early form of mastication. Later, the central cusp was shifted to one side to form a triangle. This allowed even greater occlusal surfaces for grinding and a more advanced form of mastication. The cheek teeth (molars and premolars) in particular often have a very complex topography. Mammalogists have created a vast array of names for the cusps, ridges, depressions, and other features of the occlusal surfaces of these teeth (Figure 2.7). While the full range of terminology is beyond the scope of this lab exercise, it is important for future mammalogists to understand the basic features of mammalian cheek teeth.

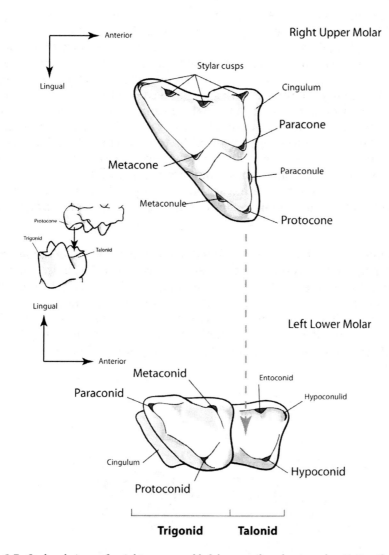

Figure 2.7 Occlusal views of a right upper and left lower tribosphenic molar. Notice that the protocone of the upper tooth fits into the basin of the talonid portion of the lower tooth.

The first rule in dental terminology is that the cusps of the upper and lower teeth are given different suffixes. The upper cusps are designated with the suffix -cone; the three main cusps are the paracone, protocone, and metacone. Secondary cusps in the upper teeth are designated by the suffix -conule (e.g. paraconule). In contrast, the cusps of lower teeth are designated with the suffix -conid, and secondary cusps use -conulid. Thus the hypoconid is a cusp in the lower molar and hypoconulid is a nearby secondary cusp (Figure 2.7). A cingulum is a shelf-like ridge around the outside of a molar (often called a cingulid for a lower molar). The stylar shelf is part of the cingulum that has become expanded and often bears several small cusps. A crista (or cristid) is a ridge and a loph is a ridge formed by the fusion or expansion of cusps.

Early therians (placentals and marsupials) had cheek teeth similar to those illustrated in Figure 2.7. The upper molars were essential triangular with three major cusps. The triangle is positioned with one side along the labial (lips) side of the tooth row and one apex pointing lingually (toward the tongue). The protocone is positioned at the lingual apex, the paracone is positioned at the anterior margin of the tooth, and the metacone is at the posterior margin. Early mammals also had a prominent stylar shelf (labial to the paracone and metacone). Because of its triangular shape, the upper cusps are said to form a trigon. Lower molars in these early mammals contained two regions; a triangular trigonid and an attached lower region called the talonid. The cusps in the trigonid are reversed (relative to those in the upper molars). Here the protoconid is on the labial side and the paraconid and metaconid on the lingual side of the jaw. The talonid contains a shallow basin surrounded by three cusps and is always on the posterior of the tooth (Figure 2.7). Because upper and lower "triangles" are reversed, the upper and lower teeth occlude (meet during chewing) precisely. The upper protocone fits snugly into the talonid basin of the lower tooth. Such tribosphenic teeth were common in early mammals, but are found in only a few modern mammals, including the Virginia opossum (*Didelphis virginianna*).

Tribosphenic teeth have been modified in a variety of ways in the lineage leading to modern mammals as their diets diversified. For example, both zalambdodont and dilambdodont teeth retain a triangular shape (Figure 2.8). In zalambdodont upper molars there is a V-shaped crest (ectoloph) with the largest cusp, the paracone, at the apex. There is also an expanded stylar shelf and the protocone is usually absent. Among living mammals, zalambdodont teeth are found in golden moles (Family Chrysochloridae) and solenodons (Family Solenodontidae).

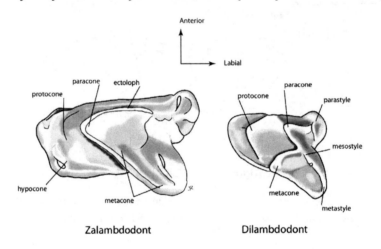

Figure 2.8 *Occlusal views of zalambdodont and dilambdodont teeth.*

Dilambdodont upper molars are similar, but have a prominent W-shaped ectoloph (Figure 2.9), with the paracone and metacone at the bottom of the W.

The ectoloph crests extend labially to the stylar shelf. Lingual to the ectoloph is the protocone. Dilambdodont molars are characteristic of an insectivorous diet and are found in living soricid shrews (Family Soricidae), moles (Family Talpidae), and many insectivorous bats (Figure 2.4).

Later in mammalian evolution a fourth cusp, the hypocone, was added lingual and posterior to the protocone (Figure 2.10). This additional cusp squared the molar forming a quadrate molar (also known as quadritubercular). Quadrate teeth are found in many mammals including hedgehogs (Family Erinaceidae), raccoons

(Family Procyonidae), and many primates (e.g., Family Hominidae). Additional modifications, such as cross-lophs, occurred throughout the evolutionary history of mammals.

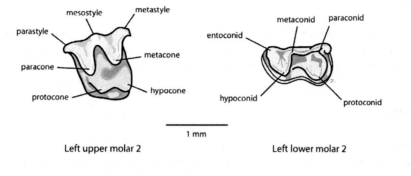

Myotis lucifugus

Figure 2.9 *Occlusal views of the upper and lower molars of a little brown bat,* Myotis lucifugus.

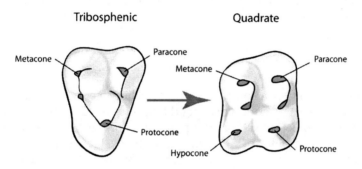

Figure 2.10 *The transition from a tribosphenic to a quadrate upper molar by the addition of a hypocone.*.

TYPES OF TEETH AND DIET

Mammals feed on a wide variety of foods. Those that feed primarily on soft, less-abrasive materials (e.g. meat) have low-crowned teeth, termed **brachydont**. In contrast, herbivores that feed on grasses and other abrasive foods have **hypsodont** (high-crowned) teeth. These teeth are subject to greater wear and therefore extend well above the gum-line. In some species, these hypsodont teeth are ever-growing.

CARNIVORES

Carnivores slice meat using blade-like cheek teeth (Figure 2.3) referred to as **secodont** teeth. The fourth upper premolar and first lower molar in the jaws form a pair of **carnassial** teeth used to slice meat (not all members of the Carnivora have carnassials). Many carnivores also have large canines for stabbing and holding prey.

PISCIVORES

Piscivores eat a diet primarily of fish. Their teeth tend to be simple and conical in shape (secondarily homodont). This type of dentition is found in several species of seals and in odontocete cetaceans (Figure 2.6).

INSECTIVORES

Insectivorous mammals eat hard and/or soft-bodied insects and other invertebrates. Their teeth are typically brachydont and either zalambdodont or dilambdodont (Figures 2.4 and 2.5). The prominent V and W-shaped ectolophs form shearing/crushing zones for breaking up the hard chitinous exoskeletons of invertebrate prey. In a few species, such as armadillos, that specialize on soft-bodied prey, the teeth are reduced to simple pegs. Monotremes (Monotremata), Old World pangolins (Pholidota) and New World anteaters (Pilosa, Myrmecophagidae) have lost teeth. Instead they rely on a long, protrusible tongue covered in sticky mucus to lap up termites and ants.

OMNIVORES

Quadrate molars with rounded cusps are characteristic of many omnivorous mammals (Figure 2.10). Omnivores tend to have varied diets consisting of a mix of hard and soft foods (including both plant and animal materials). In omnivores the four primary cusps are blunt (bunodont) and the occlusal surfaces of the upper and lower molars oppose one another directly. Bunodont teeth are usually also brachydont (low crowned). Hominid primates, pigs (Family Suidae), bears (Family Ursidae), raccoons (Family Procyonidae), and a few rodents (Family Sciuridae) are all examples of omnivores with bunodont dentitions. Many of these species retain relatively well-developed canines.

HERBIVORES

Herbivores consume a diet of leaves, seeds, fruits, grasses, and other plant material. Different plants, and even different regions of the same plant, vary greatly in abrasiveness and digestibility. Therefore, it is not surprising that herbivores have evolved many types of teeth for dealing with the challenges of an herbivorous diet. Rodents and rabbits have ever-growing incisors used for gnawing (Figure 2.11).

The incisors are chisel-shaped and strongly bevelled at the tips. The anterior surface of rodent incisors is often pigmented; the enamel here is harder than the dentine on the posterior surface, resulting in differential wear and the characteristic bevelled shape of the tips. Rodents have a single incisor in each quadrant (one pair of upper and one pair of lower incisors). Rabbits have two pair of upper incisors, but the second pair is small and located behind the first pair. Gnawing wears the incisors away at the tips, but they are rootless and grow continuously throughout the animals life.

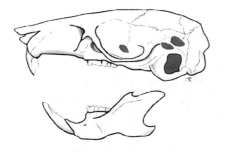

Figure 2.11 Lateral view of the skull of a deer mouse, Peromyscus, *showing the ever-growing incisors and diastema.*

Behind the incisors is the diastema, a gap where no teeth occur. The cheek teeth (premolars and molars) are typically hypsodont, flat, and have surface ridges and lophs for grinding plants. A wide variety of occlusal surfaces have evolved in rodents (Figures 2.12 & 2.13). For example, many squirrels (Sciurids) tend to have relatively simple brachydont cheek teeth, but voles and their relatives (Arvicolinae) have hypsodont teeth with complex crowns featuring folds of enamel and dentine (Figure 2.12).

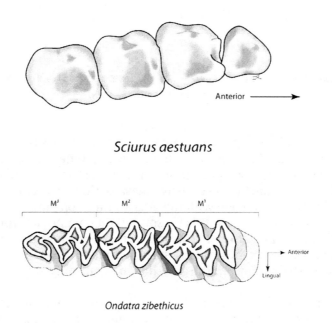

Figure 2.12   Occlusal view of the cheek teeth of a Brazilian tree squirrel, Scirurus aestuans *(top) and a muskrat,* Ondatra zibethicus *(bottom).*

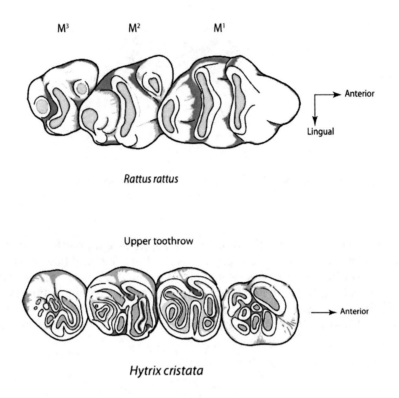

M³   M²   M¹

→ Anterior

Lingual

Rattus rattus

Upper toothrow

→ Anterior

Hytrix cristata

Figure 2.13 Occlusal view of the cheek teeth of a rat, Rattus rattus *(top) and an African crested porcu-pine,* Hystrix cristata *(bottom).*

In horses and other grazing mammals, the premolars tend to become molariform (appear molar-like). The cheek teeth have flattened surfaces with numerous exposed ridges of enamel and dentine. These ridges act to grind coarse grasses into finer particles that can be more easily digested. However, grazing animals also ingest a lot of dust and dirt along with the grasses they consume. Such an abrasive diet leads to rapid tooth wear; for this reason, grazing mammals typically have hypsodont teeth (Figure 2.14), or high-crowned teeth. Horse incisors, premolars, and molars continue to grow as the grinding surface is worn down. In an adult horse the majority of the tooth lies beneath the gum-line in deep, bony sockets (Figure 2.14).

Deer, moose, bison, antelopes, and other members of the families Cervidae and Bovidae are browsing mammals. These mammals eat leaves, soft shoots, fruits, or buds of woody plants. Browsers often lack upper incisors and canines (Figure 2.15). Instead, these teeth are replaced with a hard pad on the palate, which in concert with the lower incisors (which are generally retained) clip foliage for further processing by the cheek teeth. There is a broad gap or diastema anterior to the cheek teeth. Browsers have low-crowned (brachydont), but selenodont cheek teeth. Seleondont teeth are functionally similar to lophodont teeth (discussed below), except that the enamel ridges are oriented in an anterior-posterior plane and form characteristic crescent-shaped cusps (Figure 2.15).

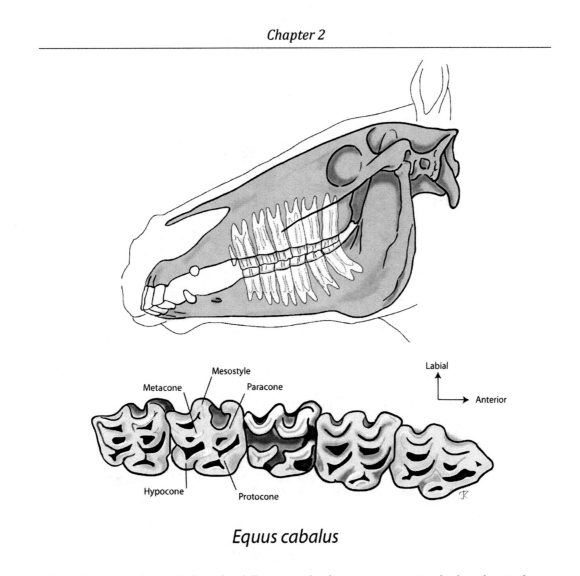

Equus cabalus

Figure 2.14 Lateral view of a horse head illustrating the elongate ever-growing cheek teeth set in deep sockets (top). Occlusal view of the right upper tooth row of a horse (bottom).

Browsing mammals may also have lophodont teeth. Lophodont teeth also have elongate ridges spanning the tooth, but in this case the lophs between cusps are perpendicular to the toothrow (labio-lingual). Many rodents (Rodentia) as well as tapirs (Family Tapiridae), and manatees (Family Trichechidae) have lophodont molars. Elephants (Family Elephantidae) and capybaras (Family Rodentia) exhibit an elaborate form of lophodonty called loxodont dentition (Figure 2.16). The lophs of loxodont teeth are numerous and form a series of closely-spaced ridges suitable for grinding hard, woody foods.

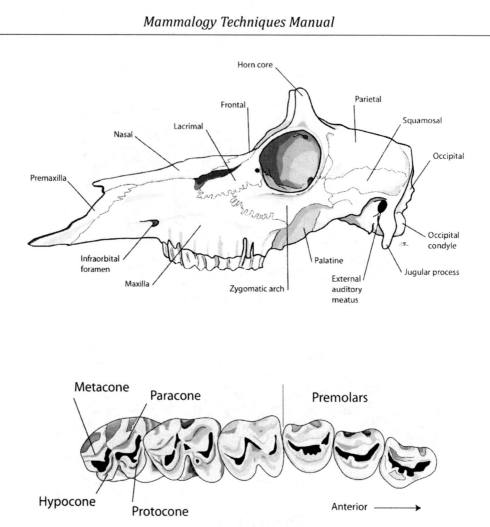

Selenodont

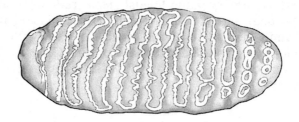

Figure 2.15 (Top) Lateral view of the skull of a pronghorn (Antilocapridae). Notice the loss of incisors and canines and the diastema anterior to the cheek teeth. (Bottom) Occlusal surface of the cheek teeth of a deer. Notice the crescent-shaped ridges that run parallel with the toothrow.

Elephas maximus

Figure 2.16 Occlusal view of the molar of an Asian elephant, Elephus maximus, *showing the many cross-lophs of a loxodont tooth.*

FILTER-FEEDERS

Although they technically lack teeth, baleen whales (Mysticete) have evolved a unique method of feeding worth mentioning here. In place of true teeth, baleen whales have racks of parallel rows of baleen in the upper jaw that serve as a sieve to filter out small crustaceans (e.g. krill and copepods) and small fish from the vast quantities of seawater they engulf during each feeding bout. Baleen is made of keratin (as are claws and hairs) and forms flat, flexible plates between 0.5 and 3.5 meters long (Figure 2.17). The distal end of each baleen plate is frayed. Two parallel rows of closely-spaced baleen plates run along the length of the upper jaw (see Vaughan et al., 2010 for details). The comb-like rows of baleen form a sieve or filter upon which to capture tiny prey.

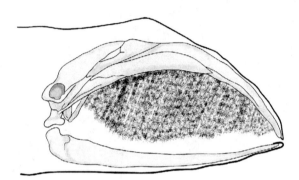

Figure 2.17   Lateral view of the skull of a baleen whale with the baleen plates intact.

Different species of baleen whales feed in different manners (see Vaughan et al., 2010), but they share a common means of filtering prey from the water. For example, humpback whales (*Megaptera novaeangliae*) swim upward through dense swarms of prey. As they near the surface, they open they mouths. The furrowed skin on the throat expands to form a vast "net" or scoop. As they swim through the school of prey, they engulf large quantities of prey just as they breach the surface (called lunge feeding). The whales closes its mouth and uses its tongue and the sieve like baleen to force out the water, trapping the tiny prey against the rows of baleen. After the water is removed, the whale then swallows its meal (Figure 2.18).

TOOTH REPLACEMENT

Reptiles, and presumably the early ancestors of mammals, have polyphyodont tooth or continuous tooth replacement. As a tooth is lost, it is eventually replaced by another tooth, and the process continues throughout the animal's life. This leads to an uneven toothrow. Mammals have two sets of teeth; a deciduous set (milk teeth) are replaced by a permanent set of adult teeth - diphyodont replacement. The milk teeth include incisors, canines, and premolars, but molars are part of the adult dentition and are not replaced. The main reason for reducing the number of lifetime tooth replacements is because mammals need a full, even tooth row with teeth that are firmly attached in order to withstand the forces required for chewing and grinding food. Among mammals, timing and style of replacement varies. For example, odontocete cetaceans have homodont dentition and only one set of teeth is ever formed; these teeth must last a lifetime. In many rodents (and some pinnipeds), replacement of the deciduous teeth takes place before birth (in utereo) and the young are born with their adult teeth.

Figure 2.18 Drawing of a humpback whale as it swims upward through a shoal of prey.

Kangaroos also have an unusual pattern of tooth replacement (Dawson, 1995). Kangaroos eat grass and other abrasive plant material and their teeth are subject to considerable wear. Placental herbivores have teeth that grow continuously as they wear, but kangaroos do not have open-rooted, continuously growing cheek teeth. Instead they solve the problem by moving a new set of molars into grinding position along the toothrow as the previous set is worn away. Thus, the molars erupt in succession from back to front as the animal ages.

Manatees are also an exception to diphyodont dentition in mammals (Domning and Hayek, 1984). They only have molars and these molars are replaced in progression from back to front. The so-called "marching molars" enter at the back of the toothrow and constantly move forward in the jaw. By the time the molar gets to the front it is well-worn and the roots degenerate, leading to the loss of the remnant of the tooth. Thus, a manatee has an indeterminate number of molars with up to seven molars in various states of wear in its jaw at any point in time.

Elephants (related to manatees in the Order Proboscidea) also have molars that get replaced throughout life, but they have only a limited set of these replacement molars (up to six per quadrant). This process is described in detail in Kingdon (1979) and is summarized below and illustrated in Figure 2.19. African elephants (*Loxodonta africana*) have up to six molars in their lower jaw in a lifetime, but only one or one and a half molars are ever present at the same time. Each molar erupts at the back of the jaw and moves forward with time. As the surface is exposed and worn away against the abrasive browse that elephants eat, the tooth is pushed further forward by those from behind. As the tooth reaches the front, it is heavily worn and the roots begin to be reabsorbed. The eventually tooth falls out (or is swallowed). If the elephant reaches 40 years of age it typically has its last (sixth) molar in place in the lower toothrow.

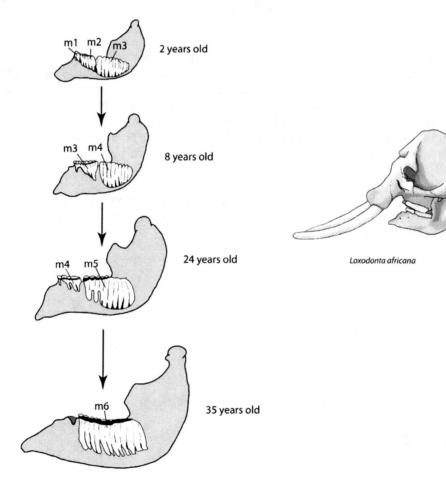

Figure 2.19 *Lateral views of the skull (right) lower jaw and molar teeth (left) in the African elephant* (Loxodonta africana). *The tusks in the skull on the right are modified incisors. The progression of jaws on the left show the movement of molars forward in the jaw as the animal ages (adapted from Kingdon 1979).*

DENTAL FORMULA

Dental formulas are simple ways to denote the complement of teeth in a species. Mammalogists typically use four quadrants when referring to the position of teeth; right side of the upper jaw, left upper jaw, right side of the lower jaw, and left lower jaw. However, when we use a dental formula we are describing the number and type of teeth in the upper and lower tooth rows of one side of the head (i.e. right upper and right lower tooth rows). Each tooth type (incisor, canine, premolar, molar) is specified along with the number or each type. There are several ways of expressing a dental formula and it is important to understand each if you are to read the vast literature on dental evolution in mammals.

Consider the dental formula for a wolf (*Canis lupis*) shown below:

$$I\frac{3}{3}C\frac{1}{1}P\frac{4}{4}M\frac{2}{3}$$

This formula indicates that the wolf has three upper and three lower incisors followed by one upper and one lower canine, four upper and four lower premolars, and two upper and three lower molars. The letters denote the type of tooth and the number represent the number of each kind of tooth in the upper (numerator) and lower (denominator) jaws. Totalling the numerators and denominators gives 10 teeth upper teeth over 11 lower teeth or 21 teeth on one side of the head. This number is doubled to get the total number of teeth in the adult animals (e.g. 42 teeth in the wolf).

Primitive metatherians and eutherians generally had more teeth than many living species. For example the dental formulas for a representative primitive metatherian is given below:

$$I\frac{5}{4}C\frac{1}{1}P\frac{3}{3}M\frac{4}{4}=50$$

During the evolutionary history of mammals, these numbers have often been reduced. The eastern cottontail rabbit (*Sylviagus floridanus*) has a reduced number of incisors and lacks canines altogether; its dental formula is:

$$I\frac{2}{1}C\frac{0}{0}P\frac{3}{2}M\frac{3}{3}=28$$

EXERCISES

EXERCISE 1: DENTAL TERMINOLOGY

Your instructor will provide you with a set of mammal skulls with a letter code to identify each skull. For each skull complete the table below by entering the dental formula, cheek tooth type, mandibular and maxillary toothrow length (in millimeters), and a hypothesized diet.

Recall the following definitions for the measurements from Chapter 1:

Length of maxillary toothrow = distance from the anterior margin of the upper canine tooth to the posterior most margin of the last molar on the maxilla.

Length of mandibular toothrow = distance from the anterior margin of the lower canine tooth to the posterior most margin of the last molar on the mandible.

Table 2.1 Data table for mammal skulls

Skull	Dental Formula	Tooth Type (i.e. selenodont)	Mandibular Toothow (mm)	Maxillary Toothrow (mm)	Diet

EXERCISE 2: DENTAL KEY TO NORTH AMERICAN MAMMALS

Using the information provide in Table 2.1, develop a dichotomous key to the skull your instructor provided. This key should use only the terminology and information described in this chapter (i.e. dental terms, tooth features, and measurements).

BIBLIOGRAPHY

Dawson, T. J. (1995) *Kangaroos, Biology of the largest Marsupials.* Comstock Publishing Associates, Cornell University Press, Ithaca, NY

Domning, D. P. and L. A. C. Hayek. (1984) Horizontal tooth replacement in the Amazonian manatee (*Trichechus inunguis*). *Mammalia*, 48:105-128.

Kingdon, J. (1979) *East African Mammals, An Atlas of Evolution in Africa. Volume IIIB, Large Mammals.* University of Chicago Press, Chicago, IL

Vaughan, T., Ryan, J.M., and N. Czaplewski. 2010. *Mammalogy*, 5th edition. Jones & Bartlett Publishing, Sudbury, MA.

APPENDIX

Couplet	Description	Go to

3

Phylogeny Reconstruction

TIME REQUIRED

One lab period of 3-4 hours

LEVEL OF DIFFICULTY

Basic

LEARNING OBJECTIVES

To understand how phylogenies are constructed.

To understand how morphological characters are used to construct phylogenetic trees.

To understand how molecular characters are used to build phylogenetic trees.

To practice manual sequence alignment and tree construction.

To use Internet-based software tools to create and explore phylogenetic hypotheses.

EQUIPMENT REQUIRED

Computers with access to the Internet

Bosque software (available at http://bosque.udec.cl/index.php)

BACKGROUND

Phylogenetics is the study of evolutionary relatedness among groups of organisms. The data used to discover these relationships may be molecular sequencing data and/or morphological data. Phylogenetics is related to, but distinct from, taxonomy - the naming and classification of organisms. Phylogenetic analyses are essential tools for elucidating the evolutionary tree of life (Figure 3.1). Evolution produced only one true phylogenetic tree of all living organisms. It is our job as scientists to recover that one true tree using the best available methods. Evolution is a change in one or more traits in a population over time. It is, therefore, regarded as a branching process, in which populations change over time and, if they change sufficiently, they may:

· speciate into separate lineages (branches),

· come together via hybridization,

· terminate by extinction.

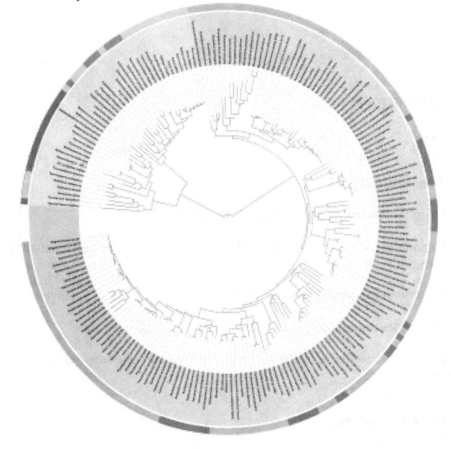

Figure 3.1 A phylogenetic tree of living organisms from the Interactive Tree of Life website.

All of these processes may be mapped on a phylogenetic tree. A phylogenetic tree is a testable hypothesis of the evolutionary events. We can never go back in time and check a tree's accuracy, but we can test the tree's branching patterns by adding more or different data - thereby testing the phylogeny. The question arises, how do we produce a phylogenetic tree (hypothesis) in the first place. The current method to infer phylogenetic trees is called cladistics. There are a number of more specific methods used in cladistics, including parsimony, maximum likelihood, and Bayesian inference. All of these require complex mathematical algorithms to establish relationships, which are then represented graphically as phylogenetic trees.

Morphological characters can be used in the reconstruction of phylogenies. However, today molecular data, which includes protein and DNA sequences, are the primary source of data for constructing phylogenetic trees. For a more complete discussion of phylogenetics see Hall and Hallgrimsson (2008) or Vaughan et al., (2010).

HOW DO WE CONSTRUCT PHYLOGENETIC TREES?

Phylogenetic trees represent patterns of ancestry and are therefore similar to family trees (genealogy). However, while families may have kept records of births and other relationships, evolution does not. It falls to biologists to reconstruct those evolutionary histories by analyzing other types of evidence, and using that data to form a phylogenetic hypothesis about how some group of organisms are related. The evidence biologists collect include morphological or genetic characters of each organism in the group they are studying. Regardless of the type of character, all characters must be heritable traits that were passed down through the lineage. The goal is to find shared derived characters that will group organisms into monophyletic clades.

STEPS IN PHYLOGENETIC RECONSTRUCTION

1. Choose a set of taxa of interest.

2. Examine each taxon carefully and determine the characters that will be included in the analysis.

3. Determine whether each character state is ancestral or derived for each taxon. This may require looking at taxa outside, but related to the group being studied (e.g. an outgroup).

4. Group taxa by shared derived characteristics (synapomorphies). Shared ancestral characters are not informative.

5. Reduce potential conflicts in the data set using parsimony (minimizing the number of conflicts).

6. Build the cladogram.

The basic steps in creating a valid phylogeny are simple and reasonably straightforward. However, the devil is in the details (as usual); constructing robust, well-supported phylogenies requires large data sets, complicated algorithms, and considerable time. To illustrate the process, we will work up a simple example using morphological characters. Suppose you are interested in the taxa in Table 3.1 and you have examined them for 10 characters. In Table 3.1 the presence of a character is denoted by an X.

Table 3.1. A table of morphological characters for selected vertebrates.

Character	Duckbilled Platypus	Iguana	Kangaroo	Chimp	Baboon	Vampire bat
Four legs	X	X	X	X	X	X
Hair	X	-	X	X	X	X
Mammary glands	X	-	X	X	X	X
Marsupium	-	-	X	-	-	-
Chorioallan-toic Placenta	-	-	-	X	X	X
Wings	-	-	-	-	-	X
Echolocation	-	-	-	-	-	X
Stereoscopic Vision	-	-	-	X	X	-
Opposable Thumbs	-	-	-	X	X	-
Viviparity	-	-	X	X	X	X

How could we use this data matrix to construct a reasonable phylogeny? The first step is to determine which character(s) are the most widely shared. In this matrix all six taxa share the presence of four legs (i.e. they are tetrapods). This character is uninformative in our quest to establish the evolutionary history of the six taxa listed because they all share it; this character must have evolved prior to the evolution of the remaining taxa. The next most widely shared characters are hair and mammary glands (shared by 5 of 6 taxa). Iguanas lack both hair and mammary glands. Therefore, iguanas must have branched off prior to the lineage containing the remaining taxa. We could illustrate this by a tree diagram like the one shown in Figure 3.2

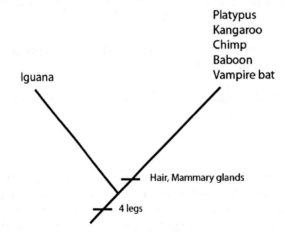

Figure 3.2 Step one in the reconstruction of the phylogeny from data in Table 3.1.

We still do not know how the platypus, kangaroo, chimp, baboon, and vampire bat are related to one another. Analysis of the next most widely shared character may help. Four of the remaining five taxa share viviparity (give birth to live young).

The platypus lays eggs so it must have branched off from the remaining taxa prior to the evolution of viviparity. The new phylogeny is now somewhat better resolved as show below in Figure 3.3.

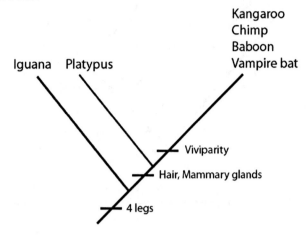

Figure 3.3   The platypus, a monotreme, branched off prior to the evolution of viviparity.

As we continue analyzing each set of shared characters, we resolve the relationships of the taxa. To continue, the next most widely shared character is having a chorioallantoic placenta. Kangaroos have a simpler choriovitelline placenta. Thus we can unite the chimps baboon, and vampire bats in a group based on the shared presence of the chorioallantoic placenta. Kangaroos must have branched from the remaining three taxa prior to the evolution of this specialized placenta (Figure 3.4).

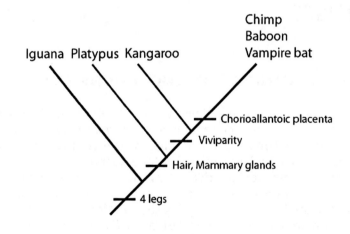

Figure 3.4   Kangaroos, members of the Metatheria, branch off next.

The next most widely shared characters are opposable thumbs and stereoscopic vision, both of which are found in chimps and baboons but not in vampire bats. There are also several characters that are unique to one taxa. These are placed on the line leading to that taxa as shown in the fully resolved tree in Figure 3.5.

LIBRARY, UNIVERSITY OF CHESTER

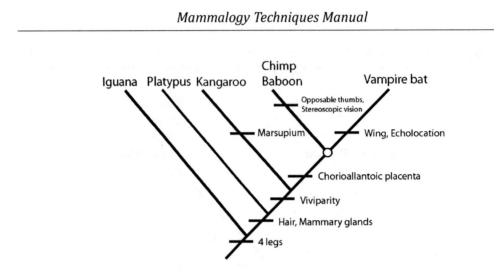

Figure 3.5   The phylogenetic tree based on the data in Table 3.1.

The taxa go on the end of the branches. Nodes are places where two or more branches come together (white dot in Figure 3.5); nodes represent the hypothetical most recent common ancestor for all taxa above that node. For example, the node in Figure 3.5 would be the most recent common ancestor of chimps, baboon, and vampire bats (eutherian mammals). A list of synapomorphies which are common to all taxa above the node are usually indicated by a horizontal bar. Of course we can also use other kinds of data, such as molecular sequence data, to construct cladograms.

EXERCISES

EXERCISE 1: MANUAL SEQUENCE ALIGNMENT

In this laboratory, you will practice two methods of determining the evolutionary relationships between organisms: molecular similarity, and cladistic analysis. Classifying based on molecular similarity uses comparisons of the sequence differences in either nucleic acids or proteins to indicate relatedness. This method has the advantage that it is relatively easy to understand. Highly related organisms are expected to have many sequence similarities; distantly related ones are expected to have low similarity. You are going to draw conclusions about the relationships of humans (*Homo sapiens*), gorilla (*Gorilla gorilla*), rats (*Rattus norvegicus*), hamster (*Mesocricetus auratus*), eagles (*Haliaeetus leucocephalus*), ostrich (*Struthio camelus*), Canada geese (*Branta canadensis*), Nile crocodiles (*Crocodylus niloticus)*, and alligators (*Alligator mississippiensis*). Obviously, these species consist of several related groups; primates form one group, rats and hamsters form a related pair (rodents), eagles, ostrich, and geese form a group of birds, and crocodiles and alligators form a pair of reptiles. What do you predict about the molecular similarity of organisms within each pair? In the first two exercises, you will use single-letter codes of the International Union of Pure and Applied Chemistry (IUPAC) to represent amino acids (Table 3.2).

Table 3.2 Single-letter IUPAC codes for the 20 standard amino acids in alphabetical order.

A alanine	G glycine	M methionine	S serine
C cysteine	H histidine	N asparagine	T threonine
D aspartic acid	I isoleucine	P proline	V valine
E glutamic acid	K lysine	Q glutamine	W tryptophan
F phenylalanine	R arginine		

The first step is a laborious one: working in groups, you must find the number of differences in amino acid sequence between each pair of species using the data in Table 3.3. There are 36 pairs of species to be analyzed, as shown by the empty cells in Table 3.4.

Table 3.3 A sample data matrix for nine taxa and the first 38 amino acids in Hemoglobin.

AA	Human	Gorilla	Rat	Hamster	Eagle	Ostrich	Goose	Crocodile	Alligator
1	V	V	V	V	V	V	V	A	A
2	H	H	H	H	H	Q	H	S	S
3	L	L	L	L	W	W	W	F	F
4	T	T	T	T	T	S	T	D	D
5	P	P	D	D	A	A	A	P	A
6	E	E	A	E	E	E	E	H	H
7	E	E	E	A	E	E	E	E	E
8	K	K	K	K	K	K	K	K	R
9	S	S	A	N	Q	Q	Q	Q	K
10	A	A	A	L	L	L	L	L	F
11	V	V	V	V	I	I	I	I	I
12	T	T	N	S	T	S	T	G	V
13	A	A	G	G	G	G	G	D	D
14	L	L	L	L	L	L	L	L	L
15	W	W	W	W	W	W	W	W	W
16	G	G	G	G	G	G	G	H	A
17	K	K	K	K	K	K	K	K	K
18	V	V	V	V	V	V	V	V	V
19	N	N	N	N	N	N	N	D	D
20	V	V	P	A	V	V	V	V	V
21	D	D	D	D	A	A	A	A	A
22	E	E	D	A	D	D	D	H	Q
23	V	V	V	V	C	C	C	C	C

AA	Human	Gorilla	Rat	Hamster	Eagle	Ostrich	Goose	Crocodile	Alligator
24	G	G	G	G	G	G	G	G	G
25	G	G	G	A	A	A	A	G	A
26	E	E	E	E	R	E	E	E	D
27	A	A	A	A	S	A	A	A	A
28	L	L	L	L	L	L	L	L	L
29	G	G	G	G	A	A	A	S	S
30	R	R	R	R	R	R	R	R	R
31	L	L	L	L	L	L	L	M	M
32	L	L	L	L	L	L	L	I	L
33	V	V	V	V	I	I	I	I	I
34	T	T	T	T	T	T	T	K	K
35	Q	Q	Q	Q	Q	Q	Q	R	R
36	F	F	Y	F	F	F	F	Y	Y
37	E	E	D	E	A	A	S	E	E
38	S	S	S	H	S	S	S	N	F

Divide the pair-wise comparisons up among group members so that each person is responsible for certain subset of pair-wise comparisons. For example, the number of amino acid differences between the human-gorilla pair is 0, and the crocodile-alligator number is 11. Enter the number of amino acid differences for each pair in the Table 3.4 (the two pair-wise comparisons just discussed are already entered for you).

Table 3.4 Table of the pair-wise amino acid differences between species for the first 38 amino acids in the hemoglobin beta chain.

	Gorilla	Rat	Hamster	Eagle	Ostrich	Goose	Croc	Alligator
Human	0							
Gorilla								
Rat								
Hamster								
Eagle								
Ostrich								
Goose								
Croc								11

1. According to the data, which two species in Table 3.4 are most closely related?

2. Why do you think these two lineages diverged the shortest time ago?

3. Which species is most closely related to humans?

4. Which is most distantly related to humans?

5. To humans, alligators and crocodiles look very similar. However, which species difference in Table 3.4 comes closest to the difference between alligators and crocodiles? How can we explain this?

EXERCISE 2: SEQUENCE ALIGNMENT USING COMPUTERS

Download and install *Bosque* software (Ramirez-Flandes and Ullo, 2008). The software can be downloaded from

http://bosque.udec.cl/downloads/BosqueSetup.exe.

Bosque runs on Linux, Windows and MacOS X (with Java runtime environment).

1. Acquire a set of amino acid sequences to use in constructing your phylogeny. To do this you will search an online database for the hemoglobin beta chain (HBB). You download the sequence data for several mammal species using Swiss-Prot proteomic server.

2. "Swiss-Prot" is one of the world's leading protein databases. Go to the Swiss-Prot database at http://us.expasy.org/sprot/. You will see the following window (Figure 3.6).

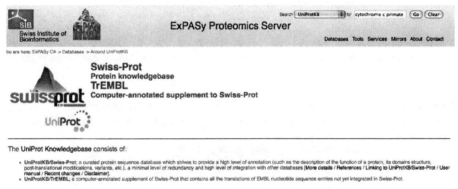

Figure 3.6   The Swiss-Prot protein database home page.

3. Using the text box at the upper right, enter the protein you want to find and the organism. For example, for primate beta chain hemoglobin, we might put in "hemoglobin primate" or "HBB primate" in the search box. Begin by searching for HBB mammal (do not use "and" or "or" in the search box). Press the *Go* button and you will see the following data table of all mammal hemoglobin beta chain sequences in the database (Figure 3.7). Each sequence is listed by its accession number along with some basic descriptive information. We only want a small subset of these sequences for our analysis.

4. Click on the small boxes to the left of each of the following species: Humans (*Homo sapiens* - #P68871), Rat (*Rattus norvegicus* - #P02091), Chimpanzee (*Pan troglodytes* - #P68873), Australian echidna (*Tachyglossus aculeatus* - #P02110), Golden hamster (*Mesocricetus auratus* - #P02094), Duck-billed platypus (*Ornithorhynchus anatinus* - #P02111), Lowland gorilla (*Gorilla gorilla* - #P02024), Common gibbon (*Hylobates lar* - #P02025), Olive baboon *(Papio anubis* - #Q9TSP1), and Squirrel monkey (*Saimiri sciureus* - #P02036). This gives us ten species. How would

you predict that these ten species will be related? Before proceeding, draw a hypothetical cladogram of the relationships between these species.

5. Select all of the species listed above by checking on box to the left of the blue link accession numbers. Click on the *Retrieve* button at the bottom of the page (in the green bar). You will see a new page (Figure 3.8) that lists your selected taxa and provides several ways to export your sequence data. You will export the data in FASTA format; this format can be readily used by a host of software.

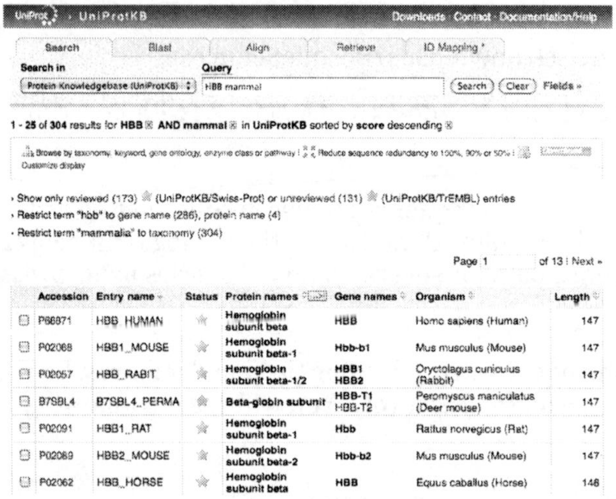

Figure 3.7   The list of all HBB sequences in the UniProtKB database for mammals.

6. Click on the *Download* link under the FASTA heading to download a text file with your data (make sure you know where it ends up on the computer - usually on the desktop or in the downloads folder).

Figure 3.8   The download page form UniProtKB.

Now you will use this sequence data to construct a phylogenetic tree for the ten species.

7. Open the *Bosque* software program (it can be downloaded at

http://bosque.udec.cl/downloads/BosqueSetup.exe.

8. The first time you open *Bosque* you will need to specify where you want to leave the file containing the local database.

9. Create a Project, by giving it any descriptive name, such as MammalsHBB. Click *Create*. Then click on the *Create Tree Project* button at the bottom left of the new window that appears (Figure 3.9).

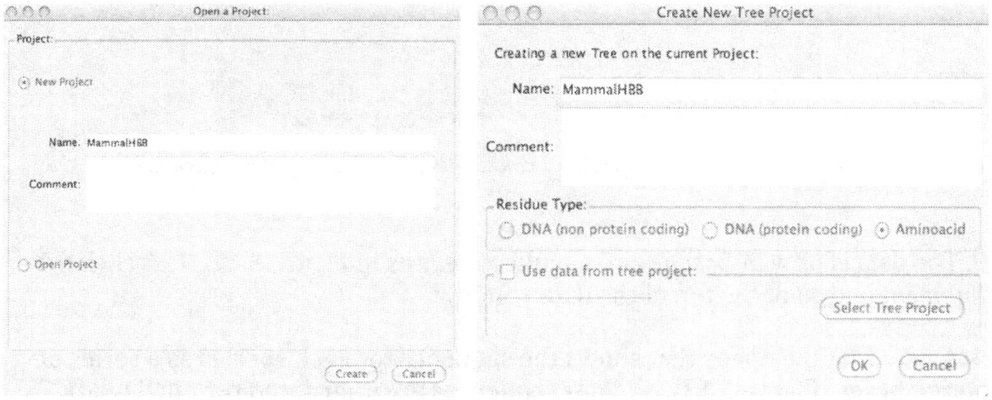

Figure 3.9   Left- The main project window in Bosque (with no sequence data imported). Right- The search screen for Bosque.

10. Create a Tree Project within your recently created Project, type the name of the project into the *Name* field as shown in Figure 3.9. Then make sure the Amino acid button is clicked (our data is for amino acids). Click *OK*. This will create a Tree Project and will present it on a new Tree Window.

11. You now have to add your set of ten sequences to the Tree Project. The *Bosque* project window looks like the one in Figure 3.10. Note that it has a row of 4 icons at the top left, a row of 10 icons in your project window (called Mammal HBB), a data field that is currently blank because you have not yet imported out FASTA sequence file.

In the top row of 4 icons, click the icon that says *Seqs* (it looks like a strand of 12. DNA). You will see a search window like the one in Figure 3.11. There are a variety of ways to use this window to acquire sequence data, but you already have your data downloaded in FASTA format. Click on the *Import from* drop-down menu at the bottom left of the screen and select *file*. This opens a new window that asks you to locate your file. Click on the *Choose file* button and locate the FASTA file on your computer and click *Open*. Then click the *import* button to bring the data into *Bosque*. A small window opens asking about tags. Click *no Tags*.

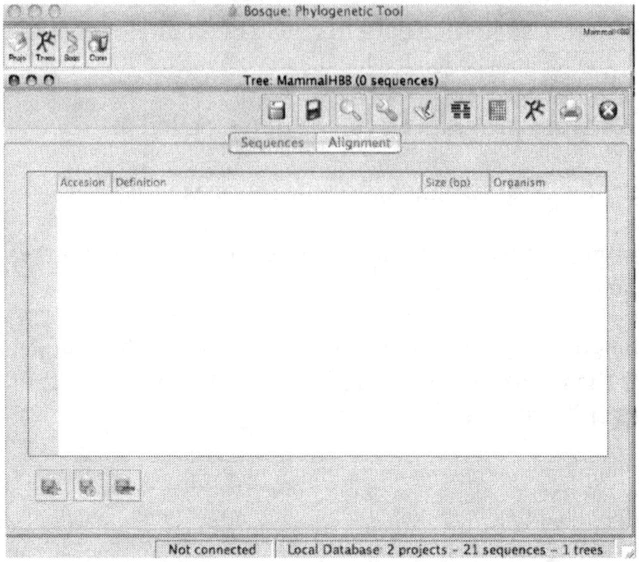

Figure 3.10 The Bosque project window without the hemoglobin data added.

13. The data is now in *Bosque* and should appear as in Figure 3.12. Next click the *Alignment* button at the top of the data window.

14. A new window opens that shows the sequence for each species as a series of colored boxes (Figure 3.13). All your sequences are unordered (or unaligned). To align them, move your mouse over the 6 icons at the bottom right of the sequence data window to show the function of each icon. Click the button *Align Sequences* at the bottom (it is the one that looks like a series of red and blue rows). A pop up window will say "Alignment by Muscle". Click *Yes* to continue, if a second pop up box appears. A new pop up window appears for the MUSCLE alignment program, click *Run*. The sequences in your *Bosque* project should now be aligned (Figure 3.13).

Figure 3.11 The Bosque search screen.

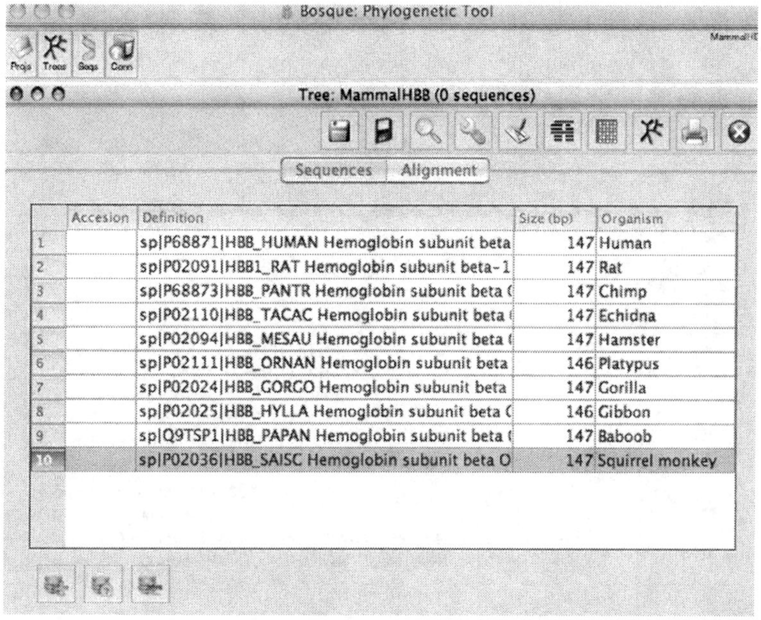

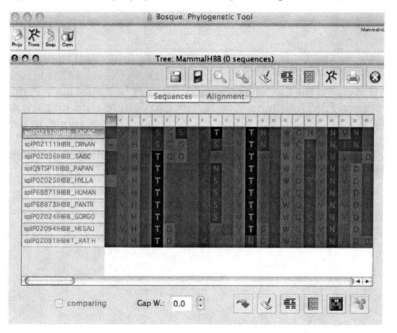

Figure 3.12   The Bosque project window with you hemoglobin data added.

Figure 3.13   Aligned sequences for hemoglobin beta chain protein for the ten species of mammals.

15. With the sequences aligned, you can now proceed to constructing a phylogenetic tree based on these sequences. To do this click on the button *Create New Tree* in the middle toolbar just above your sequence data (it looks like a green branching diagram). In the pop up that appears select *Tree by PhyML* and click *Run* in the new window that appears. When you are finished you should see a phylogeny in the project window like that in left hand part of Figure 3.14.

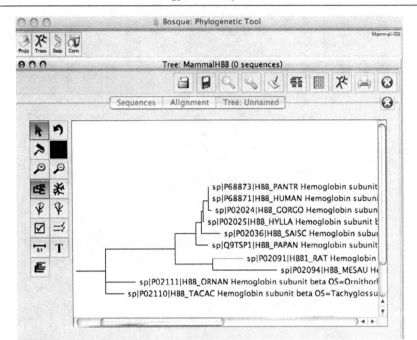

Figure 3.14 A phylogeny for ten mammal species based on the sequence data for hemoglobin beta chain proteins.

16. You have a tree on your Tree Window. The tree is difficult to read because the branch names are so long. To edit the names click on the icon that looks like a pen and paper, it is called the *Customize sequence names* button. This will open a window with a list of each FASTA sequence code. You can double-click on each line and replace the FASTA code with something much simpler. For example, in the right hand part of Figure 3.15 the scientific names were used instead of the FASTA code and the tree is much easier to read. Note that when you finish editing the names, you may have to go back and realign the sequences and then have Bosque plot a new tree.

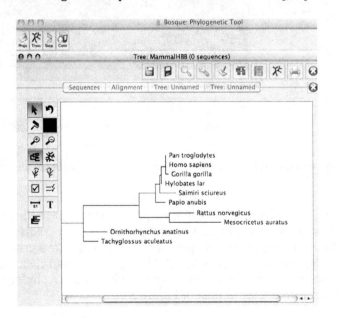

Figure 3.15 A phylogeny of mammals based on hemoglobin beta chain sequences.

17. Click on the *Compute Similarity Table* icon in the middle icon bar (it is the one that looks like a small blue data table). This gives you a similarity table much like the one you calculated by hand in Part I of this exercise (Figure 3.16).

18. The similarity table can be exported as an ASCII file with tabs or commas separating entries. This allows you to import the table into a spreadsheet program if needed.

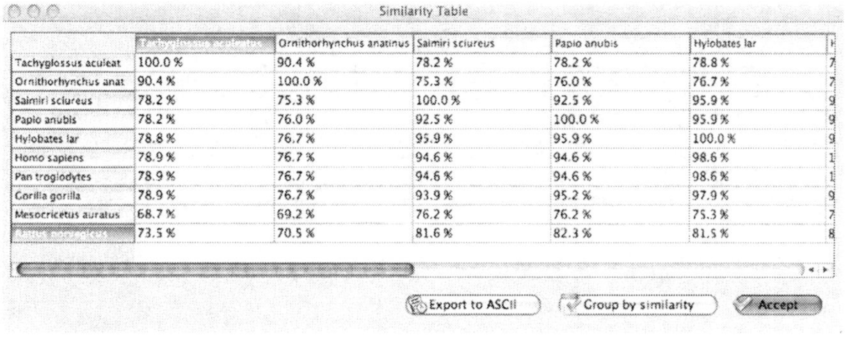

		Ornithorhynchus anatinus	Saimiri sciureus	Papio anubis	Hylobates lar	
Tachyglossus aculeat	100.0 %	90.4 %	78.2 %	78.2 %	78.8 %	7
Ornithorhynchus anat	90.4 %	100.0 %	75.3 %	76.0 %	76.7 %	7
Saimiri sciureus	78.2 %	75.3 %	100.0 %	92.5 %	95.9 %	9
Papio anubis	78.2 %	76.0 %	92.5 %	100.0 %	95.9 %	9
Hylobates lar	78.8 %	76.7 %	95.9 %	95.9 %	100.0 %	9
Homo sapiens	78.9 %	76.7 %	94.6 %	94.6 %	98.6 %	1
Pan troglodytes	78.9 %	76.7 %	94.6 %	94.6 %	98.6 %	1
Gorilla gorilla	78.9 %	76.7 %	93.9 %	95.2 %	97.9 %	9
Mesocricetus auratus	68.7 %	69.2 %	76.2 %	76.2 %	75.3 %	7
	73.5 %	70.5 %	81.6 %	82.3 %	81.5 %	8

Export to ASCII   Group by similarity   Accept

Figure 3.16   A similarity matrix for ten mammals based on hemoglobin protein.

EXERCISE 3: EXPLORING THE TREE OF LIFE

Interactive Tree Of Life is an online tool for the display and manipulation of phylogenetic trees (Letunic and Bork, 2007). It provides a host of the features and offers a novel circular tree layout. There are several pre-computed trees available for display, and you can upload and display your own data, using the 'Data upload' page. In addition you can export you tree for printing. There are also extensive help pages, for iTOL's interface.

Let's explore iTOL for mammalian phylogenies.

1. Go to the following link http://itol.embl.de/other_trees.shtml. In the NCBI taxa box near the bottom of the page, type in a taxonomic group of interest (e.g. primates) using the format Primates|subtree-genus (Figure 3.17) and click "Generate tree" button.

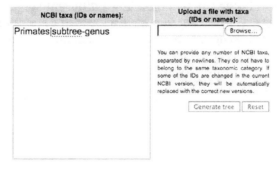

Figure 3.17   The data search interface for iTOL.

2. You will see a new page with a table of Taxonomy IDs and scientific names with "Show tree" in blue highlight. Click on the *Show tree* under the heading Scientific names. This will display a text-based description of the tree (e.g. where taxa are grouped with parentheses (Figure 3.18).

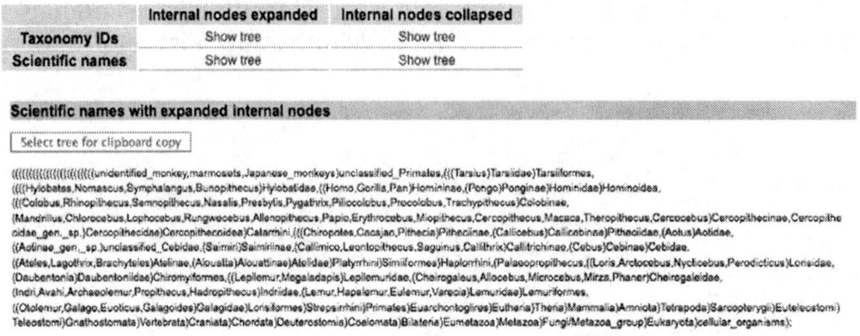

	Internal nodes expanded	Internal nodes collapsed
Taxonomy IDs	Show tree	Show tree
Scientific names	Show tree	Show tree

Scientific names with expanded internal nodes

Select tree for clipboard copy

(((((((((((((((((((((unidentified_monkey,marmosets,Japanese_monkeys)unclassified_Primates,(((Tarsius)Tarsiidae)Tarsiiformes,
(((((Hylobates,Nomascus,Symphalangus,Bunopithecus)Hylobatidae,((Homo,Gorilla,Pan)Homininae,(Pongo)Ponginae)Hominidae)Hominoidea,
(((Colobus,Rhinopithecus,Semnopithecus,Nasalis,Presbytis,Pygathrix,Piliocolobus,Procolobus,Trachypithecus)Colobinae,
(Mandrillus,Chlorocebus,Lophocebus,Rungwecebus,Allenopithecus,Papio,Erythrocebus,Miopithecus,Cercopithecus,Macaca,Theropithecus,Cercocebus)Cercopithecinae,Cercopithe
cidae_gen._sp.)Cercopithecidae)Cercopithecoidea)Catarrhini,((((Chiropotes,Cacajao,Pithecia)Pitheciinae,(Callicebus)Callicebinae)Pitheciidae,(Aotus)Aotidae,
((Aotinae_gen._sp.)unclassified_Cebidae,(Saimiri)Saimiriinae,(Callimico,Leontopithecus,Saguinus,Callithrix)Callitrichinae,(Cebus)Cebinae)Cebidae,
((Ateles,Lagothrix,Brachyteles)Atelinae,(Alouatta)Alouattinae)Atelidae)Platyrrhini)Simiiformes)Haplorhini,(Palaeopropithecus,((Loris,Arctocebus,Nycticebus,Perodicticus)Lorisidae,
(Daubentonia)Daubentoniidae)Chiromyiformes,((Lepilemur,Megaladapis)Lepilemuridae,(Cheirogaleus,Allocebus,Microcebus,Mirza,Phaner)Cheirogaleidae,
(Indri,Avahi,Archaeolemur,Propithecus,Hadropithecus)Indriidae,(Lemur,Hapalemur,Eulemur,Varecia)Lemuridae)Lemuriformes,
((Otolemur,Galago,Euoticus,Galagoides)Galagidae)Lorisiformes)Strepsirrhini)Primates)Euarchontoglires)Eutheria)Theria)Mammalia)Amniota)Tetrapoda)Sarcopterygii)Euteleostomi)
Teleostomi)Gnathostomata)Vertebrata)Craniata)Chordata)Deuterostomia)Coelomata)Bilateria)Eumetazoa)Metazoa)Fungi/Metazoa_group)Eukaryota)cellular_organisms);

Figure 3.18   iTOL screen with the scientific names.

3. Click on the *Select tree for clipboard copy* button above the data set and copy the entire entry.

4. Click on the *DATA UPLOAD* tab at the very top of the page to open a new data entry page where we will paste our tree data.

5. Paste the data you just copied into the large data box that is labeled *Paste or type the tree:* and type in a short name for the tree in the *Tree name* box. Click the *Upload* button.

6. The new page gives you some basic information about your tree. For example, it has 138 nodes and 76 leaves. You will also see an internal ID number assigned to your tree. In the "What now" section locate the line that reads, "Looks like your tree has only numbers as leaf IDs". If these are NCBI species taxonomy IDs, you can try to automatically assign taxonomic names to the internal nodes and leaves. Click on the *blue link* within that line that reads *automatically assign taxonomic names*. When the program is finished, click on the *go to main display page* link to view your tree. It should look like the one in Figure 3.19.

Notice that you can export the figure by clicking the *Export* button in the upper left.

Practice using this informative site by uploading a different subgroup of mammals and printing the phylogeny. Now that you have a basic understanding of how the *Tree of Life* site works, construct a tree of another mammal group.

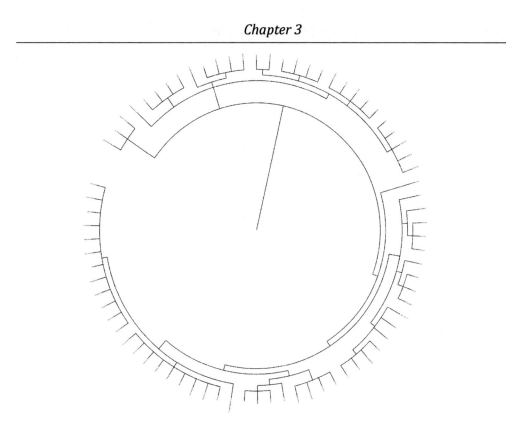

Figure 3.19 A phylogeny of Primates from the iTOL site.

BIBLIOGRAPHY

Hall, B. K. and B. Hallgrimsson (2008). *Strickberger's Evolution, The Integration of Genes, Organisms and Populations*. 4th edition, Jones & Bartlett Publishing, Sudbury, MA.

Letunic I. and P. Bork (2007). Interactive Tree Of Life (iTOL): an online tool for phylogenetic tree display and annotation. *Bioinformatics*, Jan 1;23(1):127-8. Epub 2006 Oct 18.

Ramírez-Flandes S. and O. Ullo (2008). Bosque: Integrated phylogenetic analysis software. *Bioinformatics* 24(21):2539-2541.

Vaughan, T., Ryan, J.M., and N. Czaplewski (2010). *Mammalogy*, 5th edition. Jones & Bartlett Publishing, Sudbury, MA.

4

Keeping a Field Notebook

TIME REQUIRED

One lab period of 3 to 4 hours, or over the course of several days. Ideally, students will continue to use these methods to develop a complete set of field notes for the course.

LEVEL OF DIFFICULTY

Basic

LEARNING OBJECTIVES

Learn how to record observations in a field notebook.

Understand and apply a standard format for field notes.

Hone observational skills in the field.

Understand how to read a topographic map.

EQUIPMENT REQUIRED

Field notebook

Pencils and sharpener

Topo map (of the area)

Field guide to mammals (local or regional guides)

Other field guides (optional)

Binoculars (optional)

GPS device (optional)

Digital camera (optional)

BACKGROUND

One of the most basic skills of any mammalogist is the ability to keenly observe the world and then to use those observations to draw conclusions, ask questions, or pose hypotheses. Honing your observational skills is a lifetime pursuit. Obviously, it is much easier to develop good observational skills if you already have knowledge of the species' habitat, behavior, and ecology. At first you will need to rely on other materials such as field guides and identification keys to help you characterize your observations. As you further develop your skills you will rely less and less on these tools. Field mammalogists learn a great deal by patient observation and careful note taking. General rules for good field observation include:

· Be curious about what you see.

· Slow down, be patient, and watch closely.

· Pay close attention to your surroundings: who, what, when, where, & how.

  a) who - species present.

  b) what - behaviors and habitats.

  c) when - time and date.

  d) where - location, GPS coordinates or map descriptions.

  e) how - how is the animal(s) interacting with its environment?

· Be aware of plant communities, microhabitat preferences, interactions between individuals, and any unusual activities.

· Ask questions about your observations (stick to questions that can be answered by further observations).

· Know your limits: what affects your concentration, what are your likely to overlook.

· Use available field guides, binoculars, GPS devices, cameras, local experts, etc to enhance your observations.

WHY KEEP A FIELD NOTEBOOK?

Mammalogists use field notes to record observations, remember the location and details of events, record behaviors, document the flora and fauna observed at a site, and to record data for later analysis. Our memories may fade with time, but we can always refer back to our field notes to recall certain events, dates or measurements. Important discoveries are often made when mammalogists compare recent observations with those taken in the past. For example, comparing the date of first emergence of hibernating ground squirrels with similar dates from many years ago may yield insights into the effects of climate change. Likewise, it is through careful observation and patient recording of behaviors that mammalogists have been able to decipher mating strategies, courtship behaviors, and many social interactions. It is very important to write down observations as they occur in the field and not to trust your memory and try to record them later. Field notes also allow researchers to relocate particular sites many months or years later. Many field notes include sketches or maps of the locality and the immediate environment; a map of one study site in Madagascar is illustrated in the field notes in Figure 4.1.

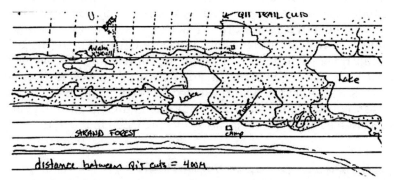

Figure 4.1   A sketch of a field site in southern Madagascar showing the relationship of trap lines to the swamp (stippled) and the base camp.

Species List: Each journal entry should include a list of species you encountered that day. It is generally a good idea to separate species by taxonomic group; put all the birds in one list and the mammals in a second listing, etc. and include scientific names when possible.

One of the most important uses of a set of field notes is to document sightings (or captures) of a species. The detailed records accompanying the sighting in the field notebook provide authenticity of the species location or behaviors; this is particularly important in mammal surveys where voucher specimens are not collected. Even if a voucher specimen is collected, field notes add critical information about habitat, weather conditions, collection methods used, and many other details that may prove useful in the future.

ELEMENTS OF A FIELD JOURNAL

Whether you follow the strict methodology, such as the Grinnell method described below, or simply keep detailed records of field trips, there are certain elements that should always appear in each journal entry. These essential elements are:

Date: The first element of each entry should include the date and, if it is important, the time of day. Dates tell us about seasonal changes as well as placing events in a chronological order. Without a date the journal entry looses a great deal of its usefulness. Generally, dates should be written using the day-month-year format. For example, 19 Dec 1989 refers to the 19th of December in the year 1989. Using the familiar US month-day-year format can get you into trouble. For example, 1/4/2010 could be January fourth 2010 to an American or it could be read as the first of April 2010 by a European. Therefore, to avoid confusion use the day-month-year format and always spell out the month. Some researchers prefer to place the date at the top of the page, others along the left margin, or with the location for each entry.

Location and Route: The location and route should be as specific as possible. Locations should begin with the broadest geographic category (usually country or state), followed by increasingly finer levels of detail. For example, in Figure 4.2 the location is given as Madagascar, Fivondronana de Tolanaro, Forest along strand, 2.5 kilometers from Manafaify (St. Luce), 24° 48' S x 47° 11' E. This location begins with the country (Madagascar), followed by the district (Tolanaro), and a specific location (2.5 kilometers from the town of Manafiafy), St. Luce is given in parentheses because this is the old French name for the town. When ever possible, the exact latitude and longitude should also be given. Today it is relatively easy to add latitude/longitude with hand held GPS devices. Nevertheless, it is important for any field biologist to know how to extract that same information from a map (batteries die at the worst possible times). For field sites within the United States it may be adequate to simply list the state, county or parish, nearest town, along with the latitude/longitude coordinates. Regardless of where you are, be precise and assume that others may want to use your notes to locate your study site in the future. Ask yourself, can they find my location easily from the location I've provided in my notes.

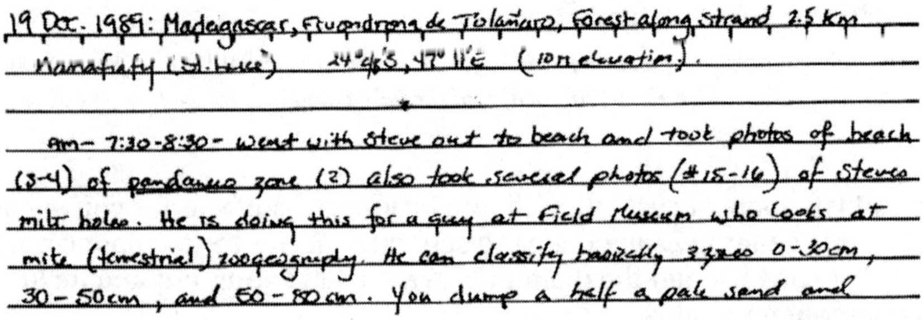
19 Dec. 1989: Madagascar, Fivondrona de Tolañaro, Forest along strand, 2.5 Km, Manafiafy (St. Luce) 24°48'S, 47° 11'E (10m elevation).

am – 7:30-8:30 – went with Steve out to beach and took photos of beach (3-4) of pandanus zone (2) also took several photos (#15-16) of Steves mite holes. He is doing this for a guy at Field Museum who looks at mite (terrestrial) zoogeography. He can classify basically 3 zones 0-30cm, 30-50cm, and 50-80cm. You dump a half a pale sand and

Figure 4.2 A typical entry from the authors field notes for a mammal survey expedition to Madagascar in 1989. Notice that the entry begins with the date and location followed by latitude/longitude and elevation.

ADDITIONAL ELEMENTS SHOULD INCLUDE:

Weather: Record general weather features such as temperature, precipitation, and cloud cover. A cool, cloudy day may yield few observations and a moonlit night may yield fewer captures.

Habitats: Record information about the plant community, forest type, and water sources near the site. These may be very important when comparing sites. Sketches or hand drawn maps may be a useful way to illustrate key elements of the habitat, locations of trap-lines, or den sites within the habitat (see Figure 4.1).

Vegetation: At times if will be useful to document the dominant plant species, as well as any fruiting/flowering species as these plants may serve as important food sites for wild mammals and birds.

Commentary: Include any other observations of unusual activity, descriptions of your collecting methods, etc. in short, descriptive sentences. The more detail you can provide the better. Try and anticipate how you will use this data in the future and attempt to provide any information that may prove useful.

TWO-PART FIELD NOTES

Some researchers prefer a two-part process for recording field notes. In this system, a small field notebook is the primary record keeper; it travels into the field, usually in a shirt pocket, and is used to record observations. Later, these field notes will be transcribed and annotated into a more formal, an larger field journal. The primary field notebook is typically small enough to fit into a pocket, leaving hands free for quick access to binoculars and other equipment. Small, inexpensive flip-top notepads are often used. Use pencils exclusively for writing field notes. Pencils write in the rain without running, and are much easier to use when sketching behavior or maps because they can create shades of light and dark gray.

The secondary field journal is a permanent record. Therefore, it contains all field observations, along with ideas and predictions derived from those observations. The field journal may be a loose-leaf notebook, a bound book, a soft-cover composition book, or even word processed file on the computer. This journal is where the original primary notes taken in the field are transcribed into a more detailed set of "permanent" notes. Each entry still contains the essential elements described above for each day. The entries also still begin with date, location, route, etc. Entries are typically written in complete, descriptive sentences. Species of interest observed in the area are listed at the end of each entry. Species collection records (with weights and measurements) are usually kept in separate section of the field journal. Many mammalogists like to keep an annual field journal, switching to a new one at the beginning of each year or field season. Others prefer a separate journal for each expedition or trip. Professionals always make back up copies (photocopies or back up disks) of their field journals and store them in a separate location from the originals.

GRINNELL METHOD

Joseph Grinnell (1877-1939), former curator of the University of California Museum of Vertebrate Zoology and influential vertebrate biologist, advocated a method that has become widely adopted (with modifications) by mammalogists around the world. Grinnell's system uses a tripartite format all integrated into the field journal; the three records are Journal, Species Account, and Catalog. For more information on Joseph Grinnell visit http://mvz.berkeley.edu/Grinnell.html.

Journal: In the Grinnell system, the field researcher uses a loose-leaf notebook to record the events and observations of the day as they occur in the field. Each entry begins with the date, time, specific location, and weather conditions placed in the upper right corner of the first page of the entry. Every time the researcher moves to a new location, a new page is used with a new heading. Within a location, observations are recorded in chronological order (often with times listed in margins). These field notes may be transcribed into a permanent field journal at a later time. Grinnell used a set of vertical and horizontal margins drawn on each page. The vertical margin was approximately 1.5 inches (3 cm) from the left edge and served to provide room for binding the loose pages into a hard cover copy at a later date (Grinnell advocated placing all such bound notes in a suitable museum or library for the use of other scientists). The horizontal margin was approximately 1.5 inches down from the top of the page. In the upper left hand corner of each page was the observer's name and the year. Only the front of each page was used for writing; the back was often used for sketching maps.

Species Accounts: Entries in this section are organized by date with a horizontal line separating each date. Below the horizontal line are a list of the species observed (or at least relevant species). Both common and scientific names are given (where possible) followed by information of the species behavior, ecology, morphology, etc.

Catalog Page: These pages are titled "Catalog" in the upper center of the page. Catalog entries refer to the specimens actually collected at the site and deposited in a museum collection for future reference. These are less common today as emphasis has shifted away from the collection of specimens. However, there are times when collection of a voucher specimen is required, or when blood or other tissues are collected non-invasively (e.g. DNA samples). A sample of a catalog data sheet is given in Figure 4.3. Because the Grinnell Method is time consuming, and requires discipline, this format has been simplified by a variety of field researchers. Nevertheless, the basic ideas presented by Grinnell as a way to document field observations are always retained.

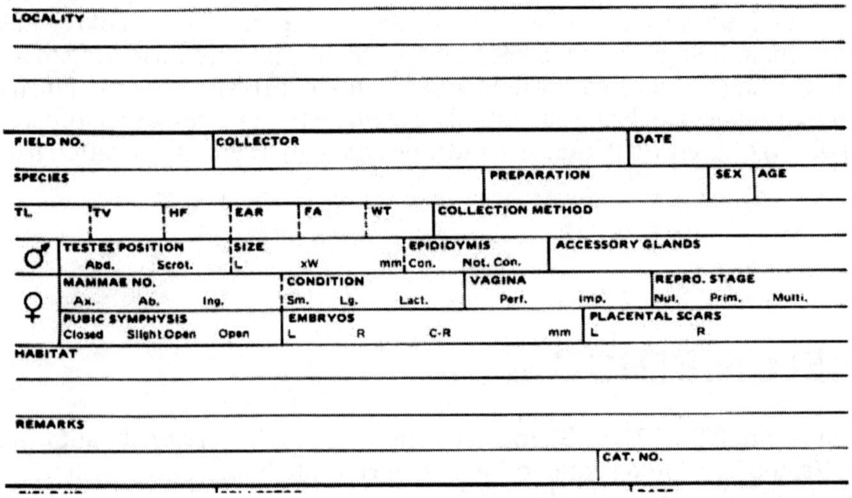
LOCALITY
FIELD NO. | COLLECTOR | DATE
SPECIES | PREPARATION | SEX | AGE
TL | TV | HF | EAR | FA | WT | COLLECTION METHOD
♂ TESTES POSITION Abd. Scrot. | SIZE L xW mm | EPIDIDYMIS Con. Not. Con. | ACCESSORY GLANDS
♀ MAMMAE NO. Ax. Ab. Ing. | CONDITION Sm. Lg. Lact. | VAGINA Perf. Imp. | REPRO. STAGE Nul. Prim. Multi.
PUBIC SYMPHYSIS Closed Slight Open Open | EMBRYOS L R C-R mm | PLACENTAL SCARS L R
HABITAT
REMARKS
CAT. NO.

Figure 4.3 A sample of part of a catalog page as used by the University of Michigan Museum of Zoology.

EXERCISES

EXERCISE 1: LOCALITY INFORMATION USING TOPO MAPS

You will be given a topo map of a study site by your instructor. Once you are at the site, you will select a suitable spot for your observations.

1. Open your field notebook and begin by entering the date and location at the top of the entry page. To do this, you will use the following basic format: a) Day,

Month, Year: Country, State, County, Township, distance and direction to nearest town (with name of town): Latitude, longitude (see instructions below)

2. Locate your study area on the topo map provided and determine the distance to nearest town or landmark.

3) Locate your exact position on the topo map and place an x at that spot in pencil on the map.

4) Use your topo map to record the latitude and longitude, and the UTM'S.

USING TOPO MAPS

First look at the margins of you topo map. There are a number of important features listed here. For example, there will be a title for your map in the upper right hand corner. This corner gives the name of quadrangle and the state where the quadrangle is located. It will also give the map series, which indicates how much land area is covered by this topo map. A map with a 7.5 minute series covers a cube of 7.5 minutes of latitude and 7.5 minutes of longitude.

In 1812, the U.S. Government created a standardized system to more accurately define a given U.S. location. This system divides land into 36 square mile units called townships. Each township has a township and range designation to define its 36 square mile area. Township is numbered north or south from a selected parallel of latitude called a base line and range, is numbered west or east of a selected meridian of longitude called a principle meridian (Figure 4.4).

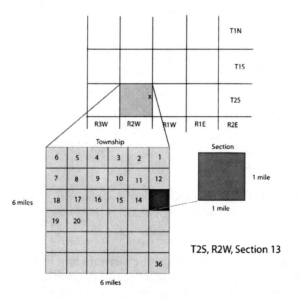

Figure 4.4   Township, range, and section information and sizes for typical topo map.

Not all US topo maps illustrate townships and range numbers. When they are used, townships are sub-divided into 36 one square mile parcels called sections. Sections are numbered from 1 to 36 for identification. For USGS topo maps section, township, and range numbers are printed in red. Section numbers are typically printed in the center of the section on the map (look for a red box with a number in it). Township numbers are printed along the right and left edge of the map (e.g. T.2S and T.3S). Range numbers are printed on the top and bottom edge of the map (e.g. R.1E and R.2E).

Map Scale: The scale on the topo map is found at the bottom center of the map. Scale bars represent a graphic scale with distance in miles, feet and meters (Figure 4.5). Use the graphic scale to estimate the distance from your site on the map to the nearest major landmark (e.g. town). The ratio scale is listed at the top of the scale bars. For example, 1:24,000, means that one inch on the map represents 24,000 inches on the ground (or 2,000 feet).

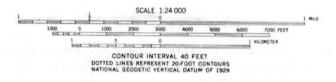

Figure 4.5 Scale bars on a typical USGS topographic map.

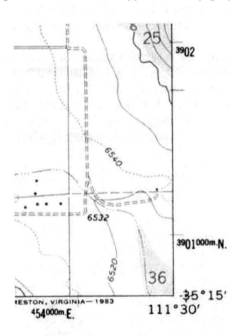

Figure 4.6   A small section of the lower right hand corner of a typical USGS topographic map showing the latitude and longitude and UTM coordinates.

Latitude/Longitude: To determine the latitude and longitude of your position, locate the numbers running all around the outside of the map (Figure 4.6). These numbers are used to represent two grid systems; the latitude and longitude system and the UTM system. The exact latitude and longitude is given at each corner of your map and at equally spaced intervals between the corners. The UTM's are the smaller bold numbers that run along the border of the map (e.g. 3901000m.N). Recall that latitude is distance measured in degrees and minutes north and south of the Equator. The Equator is 0 degrees and the poles are 90 degrees. In the northern hemisphere the latitude is always given in degrees north and in the southern hemisphere it is given in degrees south. Longitude is distance measured in degrees east and west of the Prime Meridian, which is 0 degrees longitude. As you go east or west from the prime meridian, the longitude increases to 180 degrees, and in the eastern hemisphere the longitude is given in degrees east.

Degrees are not accurate enough to find a precise location. At best, one degree of latitude and longitude would define a 70 square mile area. To get finer levels of detail, 1 degree is divided into 60' (minutes). So one minute (1') equals 1.2 miles. If even more accuracy is needed, minutes can be divide into 60 seconds (60"), where one second (1") equals 0.02 miles. The corner of the topo map in Figure 4.6 reads 35 degrees 15 minutes North latitude and 111 degrees 30 minutes West longitude.

Find you location on the topo map and run a straight line horizontally out to the right margin of the map and another vertical line to the bottom of the map. Estimate you latitude and longitude from those lines and add it to your journal heading entry. Longitude tick marks are on the top and bottom edges of the map and latitude tick marks are on the right and left edges. Note that the degrees may be left off (as an abbreviation) and you may only see the minute and/or second designations. The intersection of latitude and longitude lines may be noted by cross-marks (+).

UTM Coordinates: UTM Stands for Universal Transverse Mercator. It divides the surface of the earth up into a grid(Figure 4.7). Each grid is identified by a number across the top called the zone number (e.g. New York City is in zone 18) and a letter down the right hand side called the zone designator (e.g. Phoenix, Arizona is in UTM grid 12 S). Every spot within a zone can be defined by a coordinate system that uses meters. Your vertical position is defined in terms of meters north and your horizontal position is given as meters east. They are sometimes referred to as your northing and easting. In Figure 4.6 you can see the northing and easting coordinates on the boarder of the topo map. They are the small bold black numbers. Along the edge of the map the first UTM shown is 3901000 meters north. On a regular topo map the dash above that number would be blue. As you go up the right hand side of the map, the next UTM is 3902000 meters north. As you go up the right hand side of the map every time you pass a the small blue dash you have gone up 1000 meters. The same applies with the UTM's across the bottom of the map.

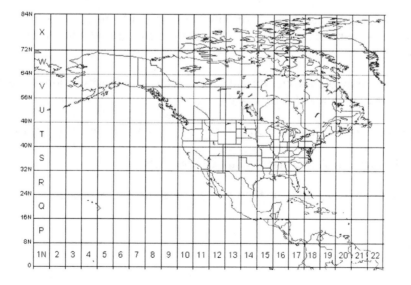

Figure 4.7 A portion of the UTM grid that covers North America.

Suppose you have a UTM position of:

18 580647.25 E BY 4504637.61 N

The first number in this UTM is 18, which is the zone number, so this map is in Zone 18. The next number is 580647.25 E (e.g. your "easting"). This means that on a topo map the position is 647.25 meters to the east of grid line 580

The final number is the "northing" and it tells you how many meters north of a particular grid line the position is. For our example, the position is 4637.61 meters north of grid line 450. (Use Google Earth to find this landmark site).

In order to figure out the UTM coordinates for your position at you study site, you need to use the map scale. If you are using a 7.5 minute topo map, then it probably has a 1:24000 scale. This ratio scale means that 1 mm measured on the topo map is equal to 24000 mm (or 24 meters) on the ground. Thus, take out a metric ruler and measure the distance in millimeters from a given UTM grid line to your position (the x you drew on the map) and multiply by 24. This is the number of meters to use in the UTM coordinate for either Easting or Northing. If you're using a 1:50000 scale map, then just multiply the number of millimeters by 50. Additional information can be found in Kjellstrom and Kjellstrom Elgin (2009) and Army (2004).

Estimate your UTM coordinates and add it to your journal heading entry. Check you estimates by verifying with a hand held GPS device provided by your instructor.

Have your instructor verify that your entry is correct.

Move to a new site and repeat the process.

EXERCISE 2: TAKING FIELD NOTES

• Find a comfortable spot, sit down and patiently begin observing and recording notes into your notebook. Be sure to note the current weather, temperature, habitats etc.

• Record all plants and animal species you can identify. If you need to look up the names in field guides (if provided).

• Spend some time recording the behaviors of one or more focal animals.

• Record questions that you are curious about.

• Turn in a copy of your field notes at the end of the session.

BIBLIOGRAPHY

Department of the Army. (2004) *U.S. Army Map Reading and Land Navigation Handbook.* Lyons Press, DE.

Kjellstrom, B. and C. Kjellstrom Elgin, (2009) *Be Expert with Map and Compass: The Complete Orienteering Handbook.* John Wiley & Sons, Hobokin, NJ.

5

Live-trapping Small Mammals

TIME REQUIRED

At a minimum this lab will require one 3-4 hour period to set up the trapping grid, followed by checking the traps twice a day for 3 consecutive days. Traps are checked in the early morning and the early evening. The time it will take to identify and process the captured animals will depend on the density of the population and the skill of the students.

LEVEL OF DIFFICULTY

Moderate to High

LEARNING OBJECTIVES

Understand the use of small mammal surveys.

Learn how to set up a standard trapping grid using Sherman live traps.

Learn how to handle small mammals.

To practice marking, measuring, and releasing small mammals.

Use the mark recapture data to estimate population size (see Chapter 6)

EQUIPMENT REQUIRED

Sherman live traps (100 traps are typically used)

Flagging

Several 50 meter tape measures

Plastic and cloth bags (to process small mammals)

Portable hair clippers (battery operated - for marking animals)

Permanent marking pen (Purple or Blue)

Pesola weight scales (60 g , 100 g , and 300 g scales are recommended)

Bait for traps

Small and large ruler for measuring body length

Backpack

Respirator masks

Latex gloves (5-6 pairs) (to handle small mammals)

Leather gloves (to handle larger mammals)

Plastic bags (to process small mammals)

Cloth bags or pillow case (to process larger mammals)

Field journal

Pencils and sharpener

Extra Sherman traps

Drinking water

BACKGROUND

Small mammals are generally defined as mammals weighing less than 5 kg as adults (Merritt, 2010). This usually includes rodents, bats, marsupials, shrews, moles, hedgehogs, elephant shrews, and a few species of small carnivores and primates (Figure 5.1). Of the more than 5,400 species of mammals, nearly 90% (i.e. 4,860 species) would be considered small mammals (Wilson and Reeder, 2005).

Understanding small mammal ecology and behavior requires a different approach than would be taken for large, diurnal mammals. Small mammals are often secretive; many species are nocturnal, semi-fossorial, or otherwise difficult to observe. Obviously, there are many questions that can only be answered by collecting detailed data on small mammal populations over long periods of time.

Examples of data requiring long term studies over multiple seasons or years include:

• home range and territory sizes

• population fluctuations over time

• effects of changing land use on small mammal communities (e.g. increasing urbanization)

• colonization patterns

• predator/prey relationships

• effects of natural disasters (e.g. floods, fires, etc.)

However, some questions can be answered by shorter, more concentrated studies. Data from such studies might include:

• species diversity in a region (e.g. Rapid Assessment Programs)

• relative abundance or density of species

• habitat preferences

• reproductive condition in a given season

• diet

*Figure 5.1 A mouse lemur (*Microcebus rufus, *Primate) captured by the author in Madagascar.*

To collect data for both long and short term studies it is often necessary to capture and handle mammals. For example, animals must be captured in order to attach radio-transmitters, ear or PIT tags (passive integrated transponders), or to collect tissue or DNA samples. One particularly useful and frequently employed method is to collect data from live-trapping studies. In this lab exercise, you will set up a standard trapping grid using Sherman live traps for small mammals, collect mark and recapture data, and estimate population size from that data.

EXERCISES

The general procedures are to:

- Set out a transect or grid array of uniquely marked live traps.

- Bait and open each trap on a specific schedule.

- Check each trap on a schedule that minimizes stress to the animals.

- Record the species, trap number, sex, and any other data deemed necessary for each animal captured.

- Mark and release each animal at the trap site where it was collected.

- Re-bait and set traps again for the next session.

Your instructor will provide additional details about how to set and bait the traps, and how to mark animals that are captured. There are risks associated with capturing and handling live mammals. It is recommended that only professionals supervise small mammal trapping and handle the mammals to avoid harm to the mammalogist and to the captured mammals.

Your instructor should review the guidelines with you prior to conducting the study. Remember, "Investigators conducting research requiring live capture of mammals assume the responsibility for using humane methods that respect target and non-target species in the habitats involved" (Gannon et al., 2007).

EXERCISE 1: SETTING UP A LIVE-TRAPPING GRID

The first step is to choose a suitable study site. Live trapping grids are generally preferred for mark-recapture studies (Figure 5.2). The size of the grid area will depend on the number of traps available and research question being asked. Generally, traps are spaced at 10 meter intervals in ten rows of ten traps each (requiring 100 traps), but smaller grids may be adequate and have the advantage of requiring fewer traps. Each trap row should contain an equal number of traps. Trap rows should also be spaced 10 meters apart.

IMPORTANT: Before conducting any live-trapping study be sure to obtain the correct permits for you state (or region). Collecting permits can usually be acquired from the state or regional conservation department (DNR, DEC, etc). In addition, it may be important to obtain approval from your institution's Institutional Animal Care and Use Committee (IACUC) even for work that occurs off-campus; consult with your institution's IACUC for advice.

Finally, It is important to follow the American Society of Mammalogists guidelines for field work with mammals (Gannon et al., 2007). The guidelines can be found on the web at: [http://www.mammal society.org/]

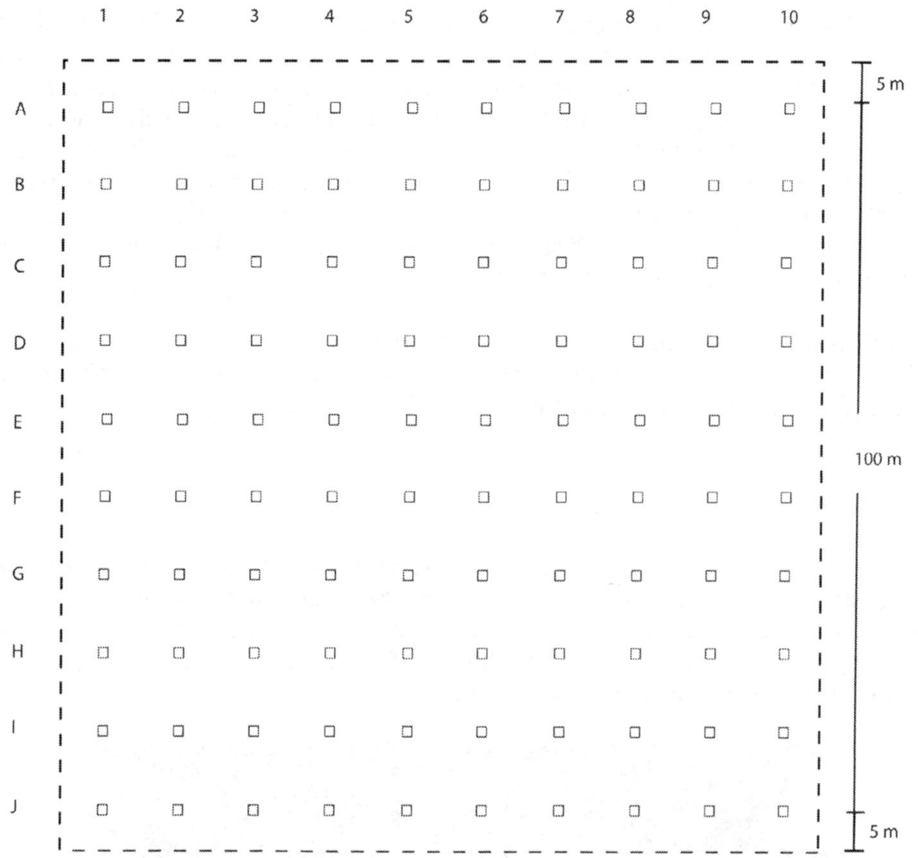

Figure 5.2   A typical trap grid of 100 trap stations (squares) laid out as a 10 by 10 grid. Note that each trap station has a unique letter and number code (e.g. A1 or H7).

1. Begin by marking off one side of the grid with a 100 meter tape (or by pacing it off).

2. Every 10 meters (or paces) along the length place a flag and label the flags A through J.

3. At each lettered flag, measure a second 100 meter line perpendicular the original line. There will be 10 such lines (one for each lettered flag) all on the same side of the original line, forming a grid.

4. Place a flag at each 10 meter interval along these perpendicular lines and label each flag with the letter of the row and a number representing the number of the 10 meter interval (Figure 5.2). Each flag records the future position of a trap on the grid and each is uniquely marked with a letter-number combination (e.g. A1, C7, B10, etc.).

5. Check the grid to make sure that the parallel lines do not converge or diverge too much. For more permanent grids in open habitats it may be necessary to use survey equipment.

6. Once the grid is established and marked, place one Sherman live trap near each flag in the grid. Place the traps so as to maximize the probability of capture; this is a real art. Experienced mammalogists look for places small mammals are likely to travel, such as along downed logs, near stumps, and within grassy clumps (Figure 5.3). In an old field, traps should not be placed on top of the grass, but within the dead grass layer close to the ground. Try and position each trap within 1 meter of the flag. In tropical regions, traps may be placed above ground along horizontal branches as well as on the ground. Be sure to record the position of each trap that is placed above the ground in your field notes. Each trap should have a small piece of painters tape (e.g. blue tape) with the letter-number code for that trap placed on the top of the trap (the tape can be easily removed during cleaning).

When the grid site or transects are to be used in future months or years, each trap location should be permanently marked or the coordinates of the corners of the grid recorded with a GPS unit so that they can be easily found again.

LIVE TRAPS

Sherman live traps (Figure 5.3) are small aluminum folding traps shaped as a rectangular box (3" x 3.5" x 9"). There are two sizes, but in most cases the larger 3" x 3.5" x 9" size works best for many small mammals. The advantage of folding traps is that they are easier to transport and clean. Information on Sherman traps can be found at: [http://www.shermantraps.com/]

Figure 5.3   A Sherman folding trap with the trap door open and set adjacent to a log.

Traps are unfolded in the field and baited with a mixture of rolled oats and peanut butter (or other comparable food items that can form a thick paste). A small ball of bait is placed directly behind the treadle (spring-plate inside trap). Some professional mammalogists prefer to wrap the bait into a small bundle using cheese cloth to prevent having to clean the traps as frequently. For capturing shrews and other insectivores, or small carnivores, it may be necessary to use a small piece of sardine or ground beef instead of oats and peanut butter. Once the bait has been place inside, the trap door is pushed down to the floor until it catches behind the small treadle plate. This holding latch can be adjusted carefully using your fingers. Attempt to set the treadle so it is just holding the door in place.

EXERCISE 2: CHECKING TRAPS AND COLLECTING CAPTURE DATA

SCHEDULE FOR CHECKING TRAPS

For a general mark-recapture survey, traps should be set for at least three nights. Check traps twice during the day, once in the early morning and again in the evening. This schedule reduces mortality, and allows both nocturnal and diurnal animals to be captured. To avoid accidental trap deaths, traps should never be set in direct sunlight or when cold snaps are forecast. If poor weather threatens, it may be necessary to add a small wad of fiberfill to the back of each trap to provide insulation until the traps are checked. Check to see that the trap's door is in the open position.

1. Traps should be checked at dawn, shortly after the sun rises the following morning. Waiting longer can lead to unnecessary animal death.

2. Each trap in the row should be carefully evaluated.

3. Check to see if the door is still open; if not, note the trap's position on the grid in your field notes. To check a closed trap, carefully lift the trap so that the front door is facing upwards (Do not shake the trap). To avoid inhalation of urine or feces keep the trap away from of your face. Slowly push open the trap door until just open enough to see if an animal is inside.

4. To remove the small mammal from the trap, hold the trap with the trap-door facing upwards, and place a large transparent plastic bag around the outside corners of the trap and over the door.

5. Slowly turn the trap and bag upside down (so the animal is on the "ceiling" and gently push open the door (with the bag still sealed around the door).

6. Slowly tilt trap until the animal slides into bag. Shake the trap lightly if needed.

7. Close the bag and weigh the bag and animal using a Pesola spring scale (Figure 5.4). Always weigh the empty bag when done and subtract the bag weight.

8. Identify the mammal to species.

9. While the animal is in the bag, gently move it toward the bottom of the bag and take a measurement of total body length and tail length using a small metric ruler.

10. While still in the bag, grab the animal by the back of the neck (through the bag), and invert the bag to expose the animal.

11. While holding it securely, mark it, determine the sex (if possible), and use a small plastic ruler to take any remaining measurements. Work quickly and methodically to avoid undue stress to the animal.

12. When marking is complete, place the bag close to the ground, open it, and wait for the animal to exit.

13. All traps should be re-baited (if needed), opened again, and replaced in their original position.

Figure 5.4   Pesola spring scales used in small mammal field research.

HANDLING CAPTURED MAMMALS

Depending on the size of the animal and the procedures required to mark or tag it, captured animals might require chemical immobilization (e.g. anesthesia) for longer handling periods. In many cases an experienced field mammalogist can remove, mark, measure, sex, and release an animal in a minute or less. However, if a radiocollar in being attached, anesthesia is generally used. It is important to remember that immobilization also causes stress (which in some cases may be greater than working efficiently without immobilization). The American Society of Mammalogists recommends that chemical immobilization be used (on a case by case basis) when animals are subject to "more than momentary or slight pain or distress" (http://grants.nih.gov/grants/ olaw/references/phspol.htm). The choice of anesthetic or analgesic and its dose and route of administration should be made in consultation with a wildlife veterinarian. (NRC, 1996).

Holding animals without causing harm to the animal (or potential bites and scratches to the researcher) takes experience. In general, handling should be kept to a minimum. Small mammals are picked up by the scruff of the neck, using thumb and first finger. The hand is rotated so the animal's belly faces the researcher and the animal's tail is grasped with the small finger . Animals over 200 g require a firmer hold on the body and isolation of the head and jaws (or anesthesia) (Figure 5.5).

CAUTION: Never hold a wild small mammal by the tail; it can cause damage to tail vertebrae and even tail loss in some tropical species. Species that are aggressive or potential disease reservoirs (e.g. *Rattus norvegicus*) should not be handled without gloves. In such cases, researchers often use a special plastic cone for immobilizing the animal. Devices to restrain rodents include conical plastic sleeves. These heavy plastic cones are transparent and have a small opening at the tip so the animal can breath. The rodent is placed in the cone head first to immobilize it (Figure 5.5). Thus restrained, it is possible to weigh, collect tail blood, or inject the animal without getting bitten.

Figure 5.5 Methods of handling larger rodents. (Left) Rodent is grasped with the thumb under the lower jaw to prevent bites. (Right) Rodent is placed head-first into a transparent plastic cone.

MARKING MAMMALS

Individual marking techniques should be

1) non-invasive,

2) allow the researcher to uniquely mark many individuals,

3) allow identification over a long period, and

4) allow marking by one person.

There are many methods for marking small mammals. Traditional methods include toe-clipping or attaching tiny numbered ear-tags. Newer, less-invasive methods include inserting PIT tags underneath the skin, or clipping or dying patches of hair. Each method has its advantages and drawbacks. For example, PIT tags can cause injury at the injection site and are very expensive.

One method that has been successful for shorter term studies is hair clipping or dying. Hair clipping involves cutting unique patterns of lines into the fur on the animals flanks. The design and position of each marked animal is recorded on the data sheet or a photograph is taken to allow future identification of that individual. Clipping is accomplished using battery powered grooming clippers (Figure 5.6). Dying techniques are also generally for short term studies. Sharpie permanent markers can be used to write a number (or letter) on the animals belly. Use blue or purple color markers as other colors are harder to read or may be confused with blood or dirt. Remember, fur must be dry or the ink will smear.

Figure 5.6   A portable, battery-operated hair clipper for use in the field.

SEXING AND AGING SMALL MAMMALS

Sexing small mammals can be difficult. Some groups (e.g. shrews and some marsupials) are difficult to sex without dissection, because the testes remain in the body even when the males are sexually active. Furthermore, in mammals both sexes have nipples although they are less prominent in males. Adult male rodents will have testes in the scrotal sacs during the breeding season, but outside of the breeding season the testes are in the abdomen (recrudescence) making sexing difficult during the nonbreeding season.

The best method of sexing rodents in the field is to note the relative distance between the anus and the penis (males) and vaginal opening (females). As illustrated in Figure 5.7, the distance is greater in males than in females.

Once the sex has been established, it is often useful to record the reproductive status of the individual (Gurnell and Flowerdew, 2006). As previously mentioned, males are considered reproductively active if their testes have descended into the scrotal sac. Reproductively active adult females are distinguished by:

· presence of perforate vagina (a small opening is present)

· presence of seminal plug in vagina (males produce seminal plugs during copulation, but these plugs are transitory and are not produced by all species)

· distended abdomen (gentle palpation of the abdomen may indicate a pregnant female)

· presence of milk in nipples (lactation)

· presence of halos (bald patches around the nipples from sucking action)

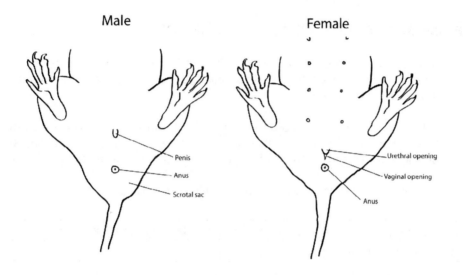

Figure 5.7 Diagrams of the ventral abdomens of a male (left) and female (right) rodent illustrating the relative positions of the genitalia and anus.

It is not possible to exactly age most live small mammals. Therefore, they are divided into age categories; neonate (or young), juvenile, and adult. Young are smaller, lack fur or have softer fur, and proportionately large heads. Juveniles have the same proportions as adults, but tend to have grayer pelage and are smaller than adults. Adults are those which are capable of reproducing.

MEASURING SMALL MAMMALS

Animals should be weighed and measured as soon as possible. Weights should be made to the nearest half gram (any greater accuracy is not needed and, considering the scale on spring balances, unlikely to be real). The standard measurements are:

total length (TL) from the distal tip of the last tail vertebrae to the tip of the nose.

tail length (T) from the distal tip of the last tail vertebrae to the vertex where the tail meets the body (tail is bent 90 degrees to find the vertex)

hind foot length (HF) from the back edge of the heel to the tip of the longest digit including the claw.

greatest ear length (E) from the notch at the base of the ear cavity to the tip of the pinna.

body weight (W) is measured (in grams) using a spring scale (e.g. Pesola) to the nearest 0.5 grams.

EXERCISE 3: DATA COLLECTION AND ANALYSIS

Record all your recapture data on the data sheet provided as well as in your field notebook. Combine all of the data from all students and trapping sessions into a master data table for the study.

Part 1: Using the data collected for you grid trapping period, create a summary table of the total number of individuals captured for each species and the trap success (percentage per trapnight for each species). Use the format in sample Table 5.1.

Table 5.1 Hypothetical summary table for a grid trapping session.

Scientific Name	Common Name	Number Captured	Percent Captured
Peromyscus maniculatus	deer mouse	296	65.8
Microtus pennsylvanicus	meadow vole	142	31.6
Sorex cinereus	masked shrew	12	2.6
	TOTALS	450	100

Part 2: Calculate the total number of trap nights (effort). Trap-nights = number of traps open per 24 hour period x number of periods (24 hours).

Part 3: Calculate the cumulative number of captures for each species over the entire trapping period (Figure 5.8).

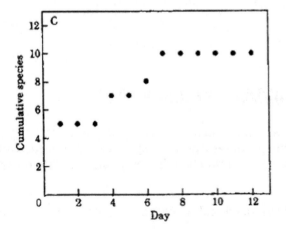

Figure 5.8 Graph of the cumulative captures for one species of small mammal over a 12 day period.

Using the data from you trapping session estimate the density of each small mammal species on the grid. First, calculate the number of unique individuals or each species on the grid. Divide the number of each species by the total area of the grid to get an estimate of density of small mammals on the study site. Add a boundary strip to the area of your grid to account for the fact that traps on the edge of the grid actually sample areas partially outside the grid. For example, if the grid was 100 meters by 100 meters with 10 meters between traps, then you would add a 5 meter (half the distance between traps) strip around the perimeter. This would mean the actual area sampled would be 110 meters by 110 meters (Figure 5.2).

BIBLIOGRAPHY

Gannon, W. L., Sikes, R. S. and ACUC (2007) Guidelines of the American Society of Mammalogists for the Use of Wild Mammals In Research. *Journal of Mammalogy*, 88:809

Grunell, J. and J. R. Flowerdew. (2006) *Live Trapping Small Mammals, A Practical Guide.* Mammal Society Occasional Publications #3, Mammal Society, United Kingdom.

Merritt, J. F. (2010) *The Biology of Small Mammals.* The Johns Hopkins University Press, Baltimore, MD.

National Research Council. (1996) *Guide for the care and use of laboratory animals.* National Academy Press, Washington, D.C.

Wilson, D. E. and D. M. Reeder. (2005) *Mammal Species of the World, A Taxonomic and Geographic Reference.* 3rd edition, The Johns Hopkins University Press, Baltimore, MD.

APPENDIX

Sherman small mammal live trapping.

DATE: _____ OBS._____ LOCATION: _____ pg ___ of ___

DAY (1-5): _____ VISIT (am/pm): _____ Start time: _____ End Time: _____ Start temp (C): _____ End temp (C): _____ %clouds: _____ Rain: _____

Fate: N=new; R=recapture; D=dead   Age: A=adult; SA=sub-adult; J=juvenile   Sex: M=male; F=female; U=unknown

Br. (breeding status): 1=non-breeding; 2=pregnant; 3=lactating; 4=testes enlarged; 5=unknown   Recap: if so, what's the mark?

Please remember to list any incidental animal sightings on the back or in the margins of this data sheet.

Trap number	Species code	Fate	Age class	Sex	Repro. condition	Recap	Mark #	Total weight (g)	Bag weight (g)	Mammal weight (g)	Total length (mm)	Hindfoot length (mm)	Ear length (mm)	Tail length (mm)	Comments

Date data entered: _____

6

Mark-Recapture Studies

TIME REQUIRED

The simulation may be completed in a typical 3-4 hour lab period.

LEVEL OF DIFFICULTY

Moderate

LEARNING OBJECTIVES

Understand the principles of mark-recapture studies.

Understand the difference between open and closed population methods.

Learn how to use the Lincoln-Petersen method

Learn how to use the Schnabel model

Learn how to use the Jolly-Seber model

EQUIPMENT REQUIRED

Plastic containers of white beans

Plastic containers of dark beans

Data sheets

Calculators

BACKGROUND

How many moose are in the Adirondack Mountains of New York State? How many elephants are in the Serengeti ecosystem? How many voles are there in a hectare of boreal forest in northern Canada? The answers to these questions require knowledge of the abundance of the mammals in question. Estimating abundance is a very common procedure for ecologists and conservation biologists. After all, population size can influence many other population parameters, including the genetic composition, rates of immigration, emigration, birth, and long term survival of the population in a given region.

Population size is determined in two ways. The first is an actual census or "head count" of all individuals within a specified region. Often this is impractical or impossible, and a second approach is to estimate population size through sub-sampling. Census techniques include aerial surveys, where the pilot flies in a grid pattern over the study area and an observer counts all individuals sighted. Aerial surveys work well for large mammals in relatively open habitats; elephants, giraffes, wild horses, and other large mammals are often counted using aerial census techniques. Obviously, direct counts of many small mammals, nocturnal or cryptic species, fossorial (burrowing) mammal, or species inhabiting densely vegetated areas are nearly impossible. In such cases, mammalogists rely on estimates of population size based upon a sub-sample of the entire population.

When direct census methods can not be used, abundance is often estimated using mark-recapture techniques (Lancia et al., 1994; Sutherland, 1996). Large mammals, such as elk or deer, are often marked with ear tags, and small mammals can be marked by tiny ear-tags, fur clipping, or by dying small patches of fur (see Chapter 5). In its simplest form, there are two session; the first session involves capturing a subset of the population, marking them, and releasing them. Marked animals are now able to move freely and intermix with unmarked members of the population. A second session involves capturing a random sample of individuals from the population and determining how many of them were marked in the first capture session. More complex techniques involve multiple recapture sessions and uniquely marking each individual. Mark-recapture techniques are based on the idea that the proportion of marked individuals in the second sample should be approximately equal to the proportion of marked animals in the total population. Thus, knowing the number of marked and unmarked animals captured in the second session, and the number originally marked in the first session, allows an estimate the total population size.

Three different mark-recapture models are presented below:

1) The Lincoln-Petersen model,

2) The Schnabel model, and

3) The Jolly-Seber model.

THE LINCOLN-PETERSEN METHOD

The Lincoln-Petersen method is the simplest because it requires only two sessions; a single marking session and a single recapture session (Seber, 1982). The data in the model include the number of individuals marked in the first sample (M); the total number of individuals that are captured in the second sample (C); and the number of individuals in the second sample that have marks from the first session (R). These data are used to estimate the total population size, N, as

$$\frac{N}{M} \approx \frac{C}{R}$$

If we are trying to estimate the population size of voles in a given area, then this equation says that the ratio of the total number of voles in the population to the total number of marked voles is approximately equal to the ratio of the number of voles in the second sample to the number of marked (recaptured) voles in the second sample. Solving for N, an estimate of the total population size yields:

$$N = \frac{CM}{R}$$

This formula is the Lincoln-Petersen index of population size. More commonly this equation is modify as:

$$N = \frac{(M+1)(C+1)}{R+1} - 1$$

The Lincoln-Petersen estimate assumes:

1. that the population is closed (no immigration or emigration and no births or deaths during the study),

2. the population does not change in size between the mark and recapture sessions,

3. the second sample is a random sample,

4. marking does not affect the recapture of individuals, and

5. marks are not lost, gained, or overlooked during the study.

LINCOLN-PETERSEN EXAMPLE:

Suppose you set up a live-trap grid of 100 Sherman traps in a 10 by 10 grid. The following morning you return to the trapping grid to check your traps, and you mark all of the individuals captured by clipping the fur on their flank. You then release each captured animal at its capture site and re-bait and re-set the traps. The next day (session two) you re-visit the grid in the morning and record the number and type (marked or unmarked) of each capture. Your data are:

M = 99

C = 149

R = 24

Entering the data into the Lincoln-Petersen equation gives:

$$N = \frac{(99+1)(149+1)}{24+1} - 1 = \frac{100(150)}{25} - 1 = 599$$

According to this, there are an estimated 599 voles at our study site. Of course, we'd like to know if this number is accurate or not. One way to determine how accurate it is (or not) is to calculate the approximate confidence limits of the estimate. To do that you need to use the following:

First, calculate p:

$$p = R/C = 24/149 = 0.16$$

Next, calculate the upper and lower confidence limits using:

$$W_1, W_2 = p \pm \left[1.96 \sqrt{\frac{p(1-p)\left(1 - \frac{R}{M}\right)}{(C-1) + \frac{1}{2C}}} \right]$$

$$W_1, W_2 = 0.16 \pm \left[1.96 \sqrt{\frac{0.16(1-0.16)\left(1 - \frac{24}{99}\right)}{(149-1) + \frac{1}{2(149)}}} \right]$$

$$W_1, W_2 = 0.16 \pm \left[1.96 \sqrt{\frac{0.16(1-0.16)(1-0.24)}{(148) + \frac{1}{298}}} \right]$$

$$W_1, W_2 = 0.16 \pm \left[1.96 \sqrt{\frac{0.16(0.84)(0.76)}{(148) + 0.00336}} \right]$$

$$W_1, W_2 = 0.16 \pm \left[1.96 \sqrt{0.102/148} \right]$$

$$W_1, W_2 = 0.16 \pm 0.0515$$

Divide M by W_1 and W_2 to give the approximate 95% confidence limits for your calculated value of N (599 voles).

$$W_1 = 100/0.21 = 476 voles$$

$$W_1 = 100/0.1085 = 922 voles$$

In this example the Lincoln-Petersen estimate gives us a vole population of 599 voles and we can be reasonably sure that the actual population is somewhere between 476 voles and 922 voles.

THE SCHNABEL MODEL

The Schnabel model is similar (in theory) to the Lincoln-Petersen method but involves more than one mark and recapture episode. This method relies on the same assumptions as before. In the Schnabel method, every time you sample the population you mark any unmarked captures and release them all back into the habitat. Thus, the proportion of marked animals in the population increases with each sampling session. When this proportion is equal to 1.0, every individual has been marked and the populations size is therefore equal to the total number marked. Obviously, you don't need to continue sampling until every individual is marked, but you need a way of modeling the change in that proportion.

SCHNABEL METHOD EXAMPLE:

Using our vole population again, assume you have the following:

S = number of sample sessions

C_i = number of animals in the *ith* sample

R_i = number of marked animals in the *ith* sample (recaptures)

U_i = number of animals marked for the first time and released in the *ith* sample

M_i = number of animals marked prior to the *ith* sample - or $\sum U_i$

Suppose you sample the trapping grid on 5 separate days (i = 5) and you have the following set of mark and recapture data (Table 6.1):

Table 6.1   *Five days of mark-recapture data used to illustrate the Schnabel method.*

i	C_i	R_i	U_i	M_i
1	10	0	10	0
2	15	5	10	10
3	10	2	8	20
4	5	0	5	28
5	5	4	1	33

The Schnabel method is essentially a series of Lincoln-Petersen samples. Here we use a modification of Schnabel called the Schumacher-Eschmeyer method:

$$N = \frac{\sum\left(C_i M_i^2\right)}{\sum R_i M_i}$$

The first step is to calculate three preliminary values which we'll just call A, B, C. They are calculated using the following formulas:

$$A = \sum C_i M_i^2$$

$$B = \sum R_i M_i$$

$$C = \sum R_i^2 / C_i$$

Using the data in the Table 6.1 we can calculate A, B, and C as (Table 6.2):

Table 6.2   *The summary values for A, B, and C calculated from values in Table 6.1*

$C_i M_i^2$	$R_i M_i$	R_i^2 / C_i
0	0	0
1,500	50	2.5
4,000	40	0.4
3,900	0	0
5,445	132	3.2
A = 14,865	B = 222	C = 6.1

Using the equation above for N, the estimate of the total population size is:

$$N = \frac{14{,}865}{222} = 67$$

The 95% confidence limits would be calculated as:

$$A / \left[B \pm t \sqrt{\frac{(AC - B^2)}{S-2}} \right]$$

where t is the Student t for 2 degrees of freedom (at 5% significance level). In the example above the 95% confidence limits would be

$$\frac{14{,}865}{222 \pm 3.182\left(\sqrt{(90{,}676.6 - 49{,}284)/3}\right)}$$

or

$$\frac{14{,}865}{\left[222 \pm 117.4\right]}$$

or 44 and 142.

Thus using the Schnabel method our new vole population is estimated to be 67 voles (or between 44 and 142 voles).

THE JOLLY-SEBER MODEL

The Jolly-Seber model does not require a closed population. Animals are assumed to be able to move in and out of the population. This method requires knowledge of survival between samples and is mathematically complex. The Jolly-Seber model includes information on when a marked individual was last captured. This requires marks or tags that are unique to the animal and records of the dates that each animal was captured. Beyond the advantage of assuming an open population, the time intervals between sampling periods do not need to be constant. This means that a population may be sampled over many months or over several years. A minimum of three sampling periods are required (more are better).

Like the models described previously, the Jolly-Seber model has several important assumptions (Krebs, 1999):

1. Every individual has the same probability of being captured in the ith sample (whether it is marked or unmarked).

2. Every marked individual has the same probability of surviving from one sampling period to the next.

3. Individuals do not lose their marks, and marks are not overlooked at capture.

4. Sampling time is negligible in relation to the intervals between samples.

THE JOLLY-SEBER MODEL EXAMPLE:

The variables required for the Jolly-Seber model are:

m_t = Number of marked animals caught in sample t

u_t = Number of unmarked animals caught in sample t

n_t = Total number of animals caught in sample t = $m_t + u_t$

s_t = Total number of animals released after sample t

m_{rt} = Number of marked animals caught in sample t, last caught in sample r

R_t = Number of the s_t individuals released at sample t and caught again in some later sample

Z_t = Number of individuals marked before sample t, not caught in sample t, but caught in some sample after sample t

The three equations are:

$$\bar{\alpha} = \frac{m_t + 1}{n_t + 1}$$

alpha is an estimate of the proportion marked at time *t*.

$$M_t = \frac{(s_t + 1)Z_t}{R_t + 1} + m_t$$

M$_t$ is an estimate of the marked population just before sample time *t*.

$$N_t = \frac{M_t}{\bar{\alpha}_t}$$

Where N$_t$ is an estimate of the population size at time *t*.

Table 6.3 *A set of hypothetical mark-recapture data for a series of capture sessions of meadow voles (Microtus pennsylvanicus).*

		Time of Capture						
		1	2	3	4	5	6	7
Time of Last Capture	1		15	1	1	0	0	0
	2			15	2	1	0	1
	3				37	2	2	0
	4					61	4	1
	5						77	3
	6							77
Total Marked (m$_t$)		0	15	16	40	64	81	82
Total Unmarked (u$_t$)		22	26	32	45	25	22	26
Total Captured (n$_t$=m$_t$+u$_t$)		22	41	48	85	89	103	108
Total Released (s$_t$)		21	41	46	82	88	99	106

In Table 6.3, the diagonal portion of data in the upper right of the table represents the actual number of marked voles collected over seven sampling session, each spaced two weeks apart. During the first capture session 22 voles were caught, marked, and 21 voles were released (e.g. one vole may have died in the trap). At sampling session two, 41 voles were caught, and of those 15 were marked in the first capture session (15 in upper box for column 2). The remaining 26 unmarked voles were marked and released at the end of capture session two. At sampling session three, 48 voles were captured. Of those 16 had marks (15 were originally marked in the last capture session two and 1 was marked last in capture session one. This illustrates an important point about the Jolly-Seber model, that marked voles may not be captured in every session; some marked voles may evade capture for one or more sessions.

Table 6.4 Highlights of the data presented in Table 6.3.

		Time of Capture						
		1	2	3	4	5	6	7
Time of Last Capture	1		15	1	1	0	0	0
	2			15	2	**1**	0	1
	3				37	**2**	2	0
	4					61	**4**	**1**
	5						77	**3**
	6							77
Total Marked (m_t)		0	15	16	40	**64**	81	82
Total Unmarked (u_t)		22	26	32	45	25	22	26
Total Captured ($n_t=m_t+u_t$)		22	41	48	85	**89**	103	108
Total Released (s_t)		21	41	46	82	**88**	99	106

In Table 6.4, take a look at the fifth sample session (column 5) 89 voles were captured, of which 64 were marked previously in either session four (61 of the marked voles) or in session three (2 of the marked voles), or in session two (1 vole). No voles from session one were recaptured in session five. Of the 89 voles captured in session five, only 88 were released (s_t). Now take a look at the horizontal row for session four (beginning with 61). This row illustrates the number of voles released at session four that were captured again in a later session. Thus, of the 61 voles marked and released at session five, four of those voles were captured again in session six, and one vole in session seven. Finally, the values in bold font represent the voles captured prior to session four, but not in session four, that were captured in a later session (e.g. for session four this would be 2 voles). Using the values in the four rows at the bottom we can calculate the following (Table 6.5):

Table 6.5 Summary data, including populations size estimates for the Microtus *population data from Table 6.3.*

Sample	Proportion Marked ($\bar{\alpha}_t$)	Size of Marked Population (M_t)	Population Estimate (N_t)
1	0.000	0.0	NA
2	0.381	19.2	50.4
3	0.347	21.8	62.9
4	0.477	47.7	100.1
5	0.722	73.1	101.2
6	0.788	87.7	111.2
7	0.761	82.0	107.7

The proportion marked in the second session is 0.381, which we calculate as:

$$\frac{M_2}{M_1 + (s_1 - m_1)}$$

or 38% of the voles in the population are marked at session 2.

The size of the marked population at session 6 is 87.7, which we calculate as:

$$M_t = \frac{(n_t + 1)Z_t}{R_t + 1} + m_t = \frac{(103 + 1)5}{77 + 1} + 81 = 87.7$$

The population size estimate at session 6 is 111.2 voles, which we calculate as:

$$N_t = \frac{M_t}{\bar{\alpha}_t} = \frac{87.7}{0.788} = 111.2$$

or 111 voles in the population at session 6.

EXERCISES

EXERCISE 1: SINGLE MARK-RECAPTURE

(LINCOLN-PETERSEN METHOD)

You can simulate a mark-recapture study in the laboratory using model organisms. Here you will use beans as model organisms. The objective of this exercise is to estimate the population size (using Lincoln-Petersen and Schnabel models, along with the 95% confidence limits). These measures are to be calculated for each trial. For the 95% confidence interval, there is a 95% certainty that the actual value for population size falls between the lower confidence limit and the upper confidence limit. Thus, more accurate predictions will have narrower confidence intervals.

To simplify the procedure for instructional purposes, you will use white beans as your model organism. Each group will have a plastic container of white beans and a container of black beans to work with. Follow the procedure below to collect your data.

1. Put 3-6 handfuls of white beans into a cloth sack (this represents the population of animals from which you will be sampling). Do not count them yet. Now make a guess as to how many beans you just placed in the sack and record this guess in the worksheet provided in the appendix.

2. Take a small handful of white beans back out of your sack. This represents your first capture session of a group (M). Count these beans and record the number as your value for M in your data table. DO NOT return these beans to your sack.

3. You will now mark the beans (organisms) you just captured. To mark these beans merely replace them with dark-colored beans. (For example, if you "captured" 25 white beans, set them aside and count out 25 dark beans to serve as your marked beans). This simulates marking them with a colored tag of some kind.

4. Now you will release the marked individuals back into the population (sack). Place the dark beans you counted out in step 3 above into the sack. The white beans that you replaced should be returned to the original white bean container (not put back into the sack). Why?

5. Shake the sack to disperse the beans in the population. Without looking, grab a small handful of beans from the sack. This represents your second capture session of a group of organisms (C). Count the total number of beans you grabbed in this handful (regardless of color) and record your answer as the value for C in your data table.

6. Examine the same handful of beans you gathered in step 5 above. Count the number of those beans that were "marked" (dark). Record this number as your value for R in the table. When you are finished counting, return this entire sample to your sack (both the white and dark beans).

7. Use the Lincoln-Petersen equation to calculate your population estimate (N). Record your answer as the value for N in your data table. Also calculate the Standard Error and 95% Confidence intervals for your estimates.

8. Now count the actual total number of beans (both white and dark) in your sack. Record your count in the worksheet.

9. Separate the light-colored beans from the dark-colored beans and return them to their original containers.

EXERCISE 2: THE SCHNABEL METHOD

1. Put 4-6 handfuls of white beans into a sack. Do not count them. Now make a guess as to how many light colored beans you just placed in the sack and record this guess in the Schnabel data worksheet in the appendix.

2. Now take a small handful of white beans back out of your sack. This represents your first capture of a group of organisms (C_1). Count these beans and record the number as your value for C_1 for trapping session 1 in your data table. This also represents the number of unmarked animals in session 1, so place this value in U_1 in the worksheet. DO NOT return these beans to your sack. Because this is the first session, there are no recaptures (R_1) or marked individuals (M_1) yet so place a 0 in each of those columns for session 1.

3. You will now mark the organisms (beans) you just captured. To mark these beans merely replace them with dark-colored beans. The number of beans you marked now becomes the number of marked individuals in the population for your next sample. Record this number of marked beans as the value for M_2 for trapping session 2 in your data table (it will be the same number as for C_1).

4. Now you will release the marked individuals back into the population (sack). Place the dark beans you counted out in step 3 above into the sack. This represents the number of marked animals prior to the next sample or M_2 in the worksheet. The white beans that you replaced with dark beans are set aside and never returned to the sack again.

5. Shake the sack.

6. Without looking, grab a small handful of beans from the sack. This represents your second capture of a group of organisms. Count the total number of beans you grabbed in this handful (regardless of color) and record your answer as your value for C_2 in your table. Also examine this handful and determine the number of marked (dark) beans and record this number as your value for R_2. Subtract R_2 from C_2 to get the number of unmarked animals in session 2 and place this value in U_2.

7. Still working with the same handful of beans collected in step 6 above, you will now mark the unmarked (white) beans in the sample by replacing them with dark beans (you are marking previously unmarked beans). Add the number of individuals you just marked (U_2) to the number marked prior to this session (M_2) and record the resulting sum as the value for M_3 for trapping time 3. This represents the total number of marked individuals now in the population. Return all the beans from this second collection (which are now all marked, and therefore dark) to the sack.

8. Shake the sack.

9. Without looking, grab a small handful of beans from the sack. This represents your third capture session. Count the total number of beans you grabbed in this handful (regardless of color) and record your answer as your value for C_3. Also examine this handful and determine the number of marked (dark) beans and record this number as your value for R_3. Subtract R_3 from C_3 to get the number of unmarked animals in session 3 and place this value in U_3.

10. Still working with the same handful of beans collected in step 9 above, you will now mark the unmarked (white) beans in the sample by replacing them with dark beans (you are marking previously unmarked beans). Add the number of individuals you just marked to the M_3 and record the resulting sum as the value for M_4 for trapping time 4. This represents the total number of marked individuals now in the population. Return all the beans from this third collection (which are now all marked, and therefore dark) to the sack.

11. Shake the sack.

12. Without looking, grab a small handful of beans from the sack. This represents your fourth capture of a group of organisms. Count the total number of beans you grabbed in this handful (regardless of color) and record your answer as your value for C_4 in your table. Also examine this handful and determine the number of marked (dark) beans and record this number as your value for R_4. Subtract R_4 from C_4 to get the number of unmarked animals in session 4 and place this value in U_4. Repeat the procedures above until you have 5 sampling sessions.

13. Calculate the population estimate (N) using the Schnabel estimate and record your answer in your data table.

14. Now count the actual total number of beans (both white and dark) in your sack. Record your count as the actual population size..

15. Separate the white beans from the dark beans and return them to their original containers.

16. Calculate the Schnabel estimate and the 95% confidence limits for the data above.

ANSWER THE FOLLOWING QUESTIONS:

1. For the simple mark-recapture in exercise I, how did your initial guess of the population size compare to the actual population number determined by a direct count?

2. For the simple mark-recapture in exercise I, how did your initial guess of the population size compare to the calculated population estimate (N)?

3. For the repeated mark-recapture in exercise II, how did your initial guess of the population size compare to the actual population number determined by a direct count?

4. For the repeated mark-recapture in exercise II, how did your initial guess of the population size compare to the calculated population estimate (N)?

5. Did the simple mark-recapture or the repeated mark-recapture provide the most accurate estimate of population size (N).

6. How accurate would these estimates be if you took a pinch of beans in your sample instead of a small handful? Why?

EXERCISE 3: THE JOLLY-SEBER MODEL USING *EXCEL*

The Jolly-Seber model can be calculated by hand but it is a complicated process if the number of sessions is large. In order to make the calculations a bit easier, you will put a set of sample data into a spreadsheet (Microsoft *Excel* or *OpenOffice*). Your instructor will show you the basic operation of a spreadsheet and how to enter formulas into a cell in a spreadsheet. Follow the procedures outlined below:

1. Enter the data for the capture histories for a small mark-recapture study of voles (*Microtus*) as shown in Table 6.6 below. Be sure to enter the data in each cell exactly as shown in Table 6.6. For example the number 14 in session 2 (see arrow) is in Cell **D5**.

Table 6.6 *A Jolly-Seber mark-recapture table for a study of* Microtus *over six trapping sessions.*

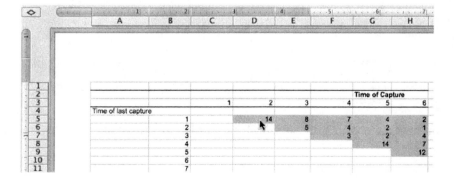

2. Below the capture histories type in the data for the first column of capture data and the number of unmarked animals in each session (u_t) as shown below.

Total marked (m_t)	0	14				
Total Unmarked (u_t)	193	25	29	38	34	21
Total Caught ($n_t=m_t+u_t$)	193					
Total Released (s_t)	193					

For the remainder of values in the Total marked (m_t) row we will enter a formula that adds the values from the session above. Click on Cell **E15** and type in the formula **=SUM(E5:E14)** and hit return. Now grab the lower left hand corner of this cell (**E15**) (the cursor turns to a +) and drag it horizontally to the right as far as cell **H15**. This will copy the SUM formula into the remaining cells in that row. It should now look like the line below.

Total Marked (m_t)	0	14	13	14	22	26

Notice that the formula adds the values in the capture history table above for each capture session (column).

3. In Cell **D17** type in the formula **=SUM(D15:D16)** and hit return. This simply adds the unmarked animals to the marked animals to get a total number of animals for that session. Again grab the lower left hand corner of cell **D17** and drag it horizontally to the right as far as cell **H17**. This will copy the SUM formula into the remaining cells in that row. Now type in the same values for s_t as you calculated for n_t. This simply means that you marked all the captures and released them all with no deaths. It should now look like the Table 6.7.

Table 6.7 *The completed Jolly-Seber data table for the* Microtus *data set.*

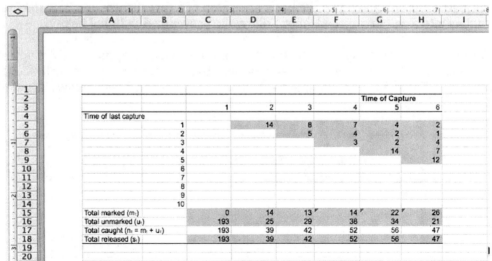

You now have your Jolly Seber table complete for this data set. Now you must calculate

The proportion marked (α_t)

Size of marked population (M_t)

Population estimate (N_t)

Probability of survival (ϕ_t)

To do this using the spreadsheet you will set up a second table below the first table. The new table will begin with Cell **A24**. Type the five column headings below into the cells listed next to each heading below.

Sample   **A24**

Proportion marked (α_t)   **B24**

Size of marked population (M_t)   **C24**

Population estimate (N_t)   **D24**

Probability of survival (ϕ_t)   **E24**

Under the Sample heading (first column) in Cells **A25** to **A30** type the numbers 1 to 6 to indicate that there are 6 sampling sessions in the study.

Now you will enter formulas for each of the remaining headings.

Recall that the formula for proportion marked (α_t) is:

$$\bar{\alpha} = \frac{m_t + 1}{n_t + 1}$$

This formula will be entered into our spreadsheet at Cell **B26**. Type 0 in Cell **B25** because there are no marked animals in the first capture session yet. In Cell **B26** we enter the formula for proportion marked (α_t) :

=(D15+1)/(D17+1)   and hit return.

Notice that is is taking M_t adding 1 and dividing that number by n_t +1. This is the same formula for the proportion marked but with references to the cell containing the data instead of the actual number.

In Cell **B27** enter the formula

=(E15+1)/(E17+1)   and hit return.

In Cell **B28** enter the formula

=(F15+1)/(F17+1)   and hit return.

In Cell **B29** enter the formula

=(G15+1)/(G17+1)   and hit return.

In Cell **B30** enter the formula

=(H15+1)/(H17+1) and hit return.

It should now look like:

Sample	Proportion Marked (α_t)
1	0.000
2	0.375
3	0.326
4	0.283
5	0.404
6	0.563

Enter the formula for the "*Size of the marked population (M_t)* in Cell **C25**. Click on Cell **C25** and enter the following formula:

=((C18+1)*SUM(D4:H4))/(SUM(D5:H5)+1)+C15   and hit return

Click on Cell **C26** and enter the following formula:

=((D18+1)*SUM(E5:H5))/(SUM(E6:H6)+1)+D15   and hit return

In Cell **C27** enter:

=((E18+1)*SUM(F5:H6))/(SUM(F7:H7)+1)+E15   and hit return

In Cell **C28** enter:

=((F18+1)*SUM(G5:H7))/(SUM(G8:H8)+1)+F15   and hit return

In Cell **C29** enter:

=((G18+1)*SUM(H5:H8))/(SUM(H9:H9)+1)+G15   and hit return

Be sure to type it exactly as it is listed above, including double parentheses. Grab the lower left corner of the Cell **C25** and drag it downward to Cell **C30** to fill in the formula for the sample sessions. It should now look like:

Sample	Proportion Marked (α_t)	Size of Marked Population (M_t)
1	0.000	0.0
2	0.375	78.6
3	0.326	99.0
4	0.283	50.1
5	0.404	83.4
6	0.563	

In Cell **D25** type NA as there is no estimate of population size with the data available from only one capture session. In Cell **D26** type in the following formula:

=C26/B26

Notice that this is the Size of the marked population (M_t) divided by Proportion marked.

In Cell **D27** type in the following formula:

=C27/B27

In Cell **D28** type in the following formula:

=C28/B28

In Cell **D29** type in the following formula:

=C29/B29

It should now look like:

Sample	Proportion Marked (α_t)	Size of Marked Population (M_t)	Population Estimate (N_t)
1	0.000	0.0	NA
2	0.375	78.6	209.6
3	0.326	99.0	304.1
4	0.283	50.1	177.1
5	0.404	83.4	206.6
6	0.563		

You could stop there and graph the changes in population size over time or you could average the population size values to get a mean estimate of the population size.

However, you might want to look at the Probability of survival for each session. This can be calculated by Clicking on Cell **E25** and entering the following formula:

=C26/(C25+(C18-C15)) and hit return.

This is equivalent to

$$\frac{M_2}{M_1 + (s_1 - m_1)}$$

In Cell **E26** and entering the following formula:

=C27/(C26+(D18-D15)) and hit return.

In Cell **E27** and entering the following formula:

=C28/(C27+(E18-E15)) and hit return.

In Cell **E28** and entering the following formula:

=C29/(C28+(F18-F15)) and hit return.

It should now look like:

Sample	Proportion Marked (α_t)	Size of Marked Population (M_t)	Population Estimate (N_t)	Probability of Survival (ϕ_t)
1	0.000	0.0	NA	0.407
2	0.375	78.6	209.6	0.955
3	0.326	99.0	304.1	0.392
4	0.283	50.1	177.1	0.946
5	0.404	83.4	206.6	0.000
6	0.563			

Save your spreadsheet. You can now enter your own data in the top table and it will automatically recalculate the estimates in the lower table for you.

Note: A detailed example of the Jolly-Seber method using the program JOLLY can also be found in Chapter 7 in this manual.

BIBLIOGRAPHY

Krebs, C. J. (1999) *Ecological Methodology*. 2nd edition, Benjamin/Cummings, Menlo Park, CA.

Lancia, R. A., Nichols, J. D., and K. H. Pollock. (1994) Estimating the number of animals in wildlife populations. pp. 215-253. In *Research and Management Techniques for Wildlife and Habitats*. 5th edition, The Wildlife Society, Bethesda, MD.

Seber, G. A. F. (1982) *The Estimation of Animal Abundance and Related Parameters*. 2nd edition, Macmillian, NY.

Sutherland, W. J. (1996) *Ecological Census Techniques, A Handbook*. Cambridge University Press, Cambridge, UK.

APPENDIX

Table 1. Lincoln-Petersen Method

Best Guess a N	
M	
C	
R	
Linclon-Petersen N	
95% Confidence Values	

Table 2. Schnabel Method

Best Guess at N	

Session	C_i	R_i	U_i	M_i
1				
2				
3				
4				
5				

$C_i M_i^2$	$R_i M_i$	R_i^2 / C_i

Schnabel N	
95% Confidence Values	

Table 3. Jolly-Seber

		Time of Capture						
		1	2	3	4	5	6	7
Time of Last Capture	1							
	2							
	3							
	4							
	5							
	6							
Total Marked (m_t)								
Total Unmarked (u_t)								
Total Captured ($n_t = m_t + u_t$)								
Total Released (s_t)								

Sample	($\tilde{\alpha}_t$)	(M_t)	(N_t)
1			
2			
3			
4			
5			
Jolly-Seber N			

7

Using CAPTURE and JOLLY Software for Mark-Recapture Data

TIME REQUIRED

One lab period of 3-4 hours

LEVEL OF DIFFICULTY

Basic to Moderate

LEARNING OBJECTIVES

Understand the basic concepts of closed and open capture models.

Introduce the concept of model fitting.

Use program CAPTURE to understand capture models: M_o, M_t, M_b, and M_h.

Understand how to compare and rank models.

Use programs JOLLY to run and understand open capture models

EQUIPMENT REQUIRED

Computers with access to the Internet

BACKGROUND

CLOSED POPULATIONS:

Recall from Chapter 6 that a closed population is one that has no additions or deletions of animals from one sampling period to the next. In other words there can not be any births or deaths, nor any movement of individuals into or out off the study area during the sampling periods. A study that is run over a short period (e.g. consecutive days) will likely meet these assumptions. In contrast, a mark-recapture study where data are collected over several months or years will likely violate the assumptions of a closed population.

In studies where the animals are abundant or the survey area is large, the resulting data set may be too large for accurately estimating population size (N) by hand (as was done in Chapter 6). Fortunately there are a number of software packages that can be used to analyze these more complex data sets. Most of these involve sophisticated statistical techniques, such as Maximum Likelihood Estimation that are beyond the scope of this discussion (but see Williams et al., 2001). We will use online versions of CAPTURE and JOLLY software to analyse several sample populations. Before we dive into the analysis, it is important to understand a few basic concepts.

CAPTURE PROBABILITY AND ENCOUNTER HISTORIES

Assume we complete a three day mark-recapture study. On the first day we capture, mark and release 55 rodents on our study grid (each animal receives a unique mark). On days two and three we capture, record marks, and release the animals without adding any new marks to the unmarked individuals (so for simplicity of this discussion we always have 55 marked animals in the population). The basic model is shown in Figure 7.1.

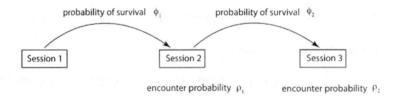

Figure 7.1   Encounter history and probability for each history.

In Figure 7.1 we have three sampling days (session 1 to 3) denoted by boxes. Time runs from left to right. The arrows between sessions denote the probability that an animal marked in session 1 will survive until sampling session 2 ($\emptyset_1$) and that, having survived until session 2, the animal will survive until session 3 ($\emptyset_2$). The probabilities below sessions 2 and 3 are the probability of encountering a marked animal (p_1 and p_2). Thus the probability of encountering any animal at a session is determined by both $\emptyset$ and p.

Recall that we only marked animals in the first session. Every time we encounter a marked animal we record the encounter with a 1. If that animal was not seen in a session it is recorded as a 0. Thus we can create an encounter history for each of the 55 animals originally marked (Figure 7.2). In Figure 7.2, the top encounter history would mean that this individual was marked in session 1 and recaptured (encountered) again in both session 2 and session 3. To be encountered it must also have survived the intervals between session 1 and 2 and session 2 and 3. Thus, the probability of recording this particular encounter history (e.g. 111) is equal to:

$$\phi_1 \, p_2 \, \phi_2 \, p_3$$

Encounter history	Probability
111	$\phi_1 \ p_2 \ \phi_2 \ p_3$
110	$\phi_1 \ p_2(1 - \phi_2 \ p_3)$
101	$\phi_1 (1 - p_2) \ \phi_2 \ p_3$

Figure 7.2 Encounter history for three animals along with the probability of that encounter history.

The second animal in Figure 7.2 has an encounter history of 110. This animal was marked in session 1 and recaptured (encountered) in session 2 but not encountered in session 3. The third animal has an encounter history of 101, corresponding to being marked in session 1, not captured in session 2, but recaptured again in session 3. Obviously, even though it was not captured in session 2 it did survive that interval.

Note: There is one encounter history that is possible but not shown, what is it and what would its probability be? Could this study yield an encounter history of 011?

Obviously, with 55 marked animals, there will be several animals that share a particular encounter history. We could list each of the 55 animals separately in a long column. Alternatively, because we know there are only 4 possible encounter histories, we can add this information to our encounter history data as:

111 7;

110 20;

101 15;

100 13;

This set of encounter histories is read as "there are 7 animals that have an encounter history of 111."

In order to use the software to calculate the probabilities and maximum likelihood estimates we need to put our mark recapture data into an encounter history format that the software can understand. The basic data format for encounter histories for each animal caught over the course of the study, is described as a matrix whose columns represent trapping sessions (time periods), and whose rows represent animals. The matrix entries are either "1" (denoting a capture) and "0" (denoting not captured). Consider the following set of encounter histories:

Table 7.1 Sample Mark-Recapture Encounter Histories.

Population A	Population B
101111	1011 2
001010	1100 10
010000	1101 1
000001	1001 4
110010	0101 3
100011	1011 12
111000	1010 8
100010	1000 10

Notice that in Table 7.1 it is possible to mark new animals as they are encountered in the study (e.g. a 0 in the first column). How many individuals are represented in the left hand matrix? How many in the right hand matrix?

THE CAPTURE MODELS

As mentioned above complex statistical models have been developed to model capture probabilities and estimate population size (N). If there are only 2 sampling periods, however, these models do not generally apply and it is best to use the Lincoln-Petersen method described in Chapter 6. If the number of sampling sessions is greater than 2, then there are a number of models available. We will look at five models that vary in certain ways. They are:

· M_0 - Neither behavioral nor temporal variation nor capture heterogeneity

· M_b - Behavioral response only

· M_t - Temporal variation only

· M_h - Individual capture heterogeneity

· M_{bh} - Behavioral response and capture heterogeneity

Let's look at each model a bit closer. The first model M_0 assumes that there is no variation in capture probability over time (between sessions) and no change in behavior of the animal toward the traps (e.g. no trap-shy or trap-happy animals). This is the simplest model because if all encounter and survival probabilities are the same, then the model looks like that in Figure 7.3.

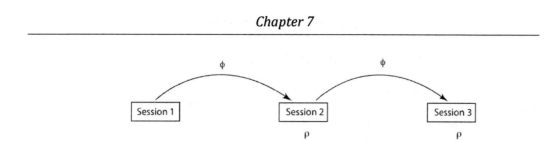

Figure 7.3   Probability history for the M_0 model.

Obviously, this is not always realistic. The behavioral model (M_b) allows changes in the behavioral response between sessions, such as a change in capture probability associated with previous capture history. For example, marked and unmarked animals can now have different encounter (capture) probabilities. This can occur when animals become trap-happy (e.g. an increased probability of being captured after the initial capture) or trap-shy (a decrease capture probability in subsequent sessions). The temporal model (M_t) assumes that each animal has the same capture probability for each trapping session, but capture probabilities vary from one session to the next. In the individual capture heterogeneity model (M_h) there is no variation in temporal or capture probabilities, but each animal may have a different capture probability. Finally, the M_{bh} model has two sources of variation, behavioral variation and potentially different capture probabilities between animals (details and mathematical equations for each can be found in Williams, et al., 2001).

EXERCISES

EXERCISE 1: USING THE PROGRAM CAPTURE

Program CAPTURE calculates estimates of capture probability and population size for closed population mark-recapture data (Rexstad and Burham, 1991; White et al., 1978). CAPUTRE generates output for:

· models M_0, M_b, M_t , M_h, and M_{bh} ,

· goodness of fit and model comparison statistics,

· estimates of abundance,

· capture probability, and SE of the estimates under each model,

· and a test of the closure assumption.

CAPTURE can be downloaded as a stand alone software package for Windows PCs from the Patuxent Software Site:

http://www.mbr-pwrc.usgs.gov/software.html

or you can use the convenient online version at:

http://www.mbr-pwrc.usgs.gov/software/capture.html.

You will use the online version in this exercise. To get started, you'll need some mark-recapture data in the format readable by CAPTURE. Type (or cut and paste) in the data in Appendix A.

The top three lines are:

```
TASK READ CAPTURES OCCASIONS=6 X MATRIX
FORMAT='(A1, 1x, 6F1.0)'
READ INPUT DATA
```

The first line tell the program CAPTURE to read a set of data in a capture matrix with 6 encounter sessions (occasions = 6). There are four ways data can be entered, but you will only learn the matrix format here (for details on the others consult White et al., 1978). The second line is a FORTRAN format statement that tells CAPTURE that the data have a label (each animal is given a unique letter), followed by a space, followed by a string of 6 numbers. In this case the 6 numbers are a string of "1s" and "0s", where a 1 represents an encounter (capture) and a 0 represents no encounter in that session. The third line tells CAPTURE to input the data that follows. Below the three commands is the data matrix itself, beginning with animal "a" with its encounter history of "100010" and ending with animal "z" and its encounter history "001111".

a 100010

b 100000

c 100010

...

z 001111

The data matrix is followed by four more command lines (tasks) that tell CAPTURE what you want it to do with the data. In your example, you are asking CAPTURE to test to see if the population is a closed population, to test the models and determine the most appropriate model to fit to this data set, and to estimate the population size. The commands are:

```
TASK CLOSURE TEST
TASK MODEL SELECTION
TASK POPULATION ESTIMATE ALL
```

The first TASK helps you determine if the assumption of a closed population is valid for your data set. The TASK MODEL SELECTION command produces a series of hypothesis tests that help determine which of the models provides the best estimate. The final command provides estimates of population size for each model.

Step 1. Type (or cut-and-paste) the lines in Appendix A into the white data entry box in the online version of the program CAPTURE,

Step 2. Click on the *perform Analysis* button at the bottom of the page. A new window opens with a lot of useful information. It is best to print (or cut-and-paste it into a text file) this output now so that you can follow along with the description of the output below.

DATA OUTPUT IN CAPTURE

This simple analysis yields about 18 pages of output. What does it all mean? The first information you see in the output file is a summary of what the output file contains. Sometimes you'll see an input error (unidentified task) listed here, but usually this is due to an extra line and can probably be ignored. If you get a lot of input errors, it means you did not type in the data correctly.

In this analysis, or output file should begin with:

```
CAPTURE output...
Mark-recapture population and density estimation program
Page 1
             Program version of 16 May 1994 16-Jun-10
Input and Errors Listing
 Input---task read captures occasions=6 x matrix
** Warning ** captures= 6.00000 assumed.
Input---format='(A1, 1x, 6F1.0)'
Input---read input data
     Summary of captures read
             Number of trapping occasions 6
             Number of animals captured 26
             Maximum x grid coordinate 1.0
             Maximum y grid coordinate 1.0
Input---task closure test
Input---task model selection
Input---task population estimate all
Input---task population estimate appropriate
```

You should confirm that this is appropriate for the data set you used. Remember, the program does what you tell it to, so if you put in garbage you get out garbage. It is the users responsibility to ensure that the analysis is preformed correctly.

The next set of output (page 2) contains the following lines:

Test for closure procedure. See this section of the Monograph for details.

```
Overall test results --
     z-value -0.508
     Probability of a smaller value 0.30570
```

The closure test suggests that the population is closed (z = -0.508, P=0.30570). If the probability was smaller (e.g. 0.00519) it would suggest the population was not closed.

Immediately below this test you will see a summary table that includes (among other things) the following summary data:

```
Model selection procedure. See this section of the Monograph
for details.
Occasion j= 1 2 3 4 5 6
Animals caught n(j)= 11 15 14 14 12 11
Total caught M(j)= 0 11 21 26 26 26
Newly caught u(j)= 11 10 5 0 0 0
Frequencies f(j)= 2 7 10 4 3 0
```

Notice that in this table the number of animals caught over the six sessions was fairly constant. This is consistent with a closed population. If, the number of captures had increased sharply in sessions 5 and 6, then it might indicate that individuals immigrated onto the trapping grid toward the end of the study. Conversely, if the number of captures dropped off sharply toward the end of the study it might indicate that some individuals either left the study area or died. In either case, the assumption of a closed population is violated. In this example, you can assume that the population is closed.

Following this summary table are Chi square statistics and goodness of fit tests for model comparisons. These are summarized in a table at the end of these tests. It should look like this:

Model	M_o	M_h	M_b	M_{bh}	M_t	M_{th}	M_{tb}	M_{tbh}
Criteria	1.00	0.82	0.46	0.75	0.00	0.42	0.36	0.78

The values for each model in this table can be roughly interpreted as the probability that the model is the "correct" model. Model M_o, having the highest value (1.00) appears to be the best model. However, use caution in just accepting the weight given in this table. For example, when data sets are small, M_o will probably be weighted 1.00 and therefore the "best" model. However, M_o is not a very robust model and it should probably not be used if there is another model with a reasonably high value. Recall that model M_o assumes that there is no behavioral or temporal variation and no capture heterogeneity. This is unlikely to be true in most studies.

In your case, you might choose to look more closely at models M_o and M_h. Below this table you should find output for each model. Find each of the models just mentioned and create a summary table of their estimates for population size, standard error, approximate 95% confidence intervals, and profile likelihood intervals.

	M_o	M_h
Population estimate	26	30
Standard error	0.7041	2.1417
95% confidence interval	26–26	28–36
Profile likelihood interval	26–28	–

Model M_h, which is second best (0.82 vs. 1.00) is far more robust, and should probably be used. The estimate of N is 30 animals with a 95% confidence interval of 28 to 36 (M_h is not likelihood based and so there are no profile likelihood intervals reported for this model).

EXERCISE 2: TIGERS IN INDIA

Go ahead and repeat this analysis, using the second set of data (in Appendix B). These data are from a survey of tigers (Karanth et al., 2004) in Kanha National Park in Madhya Pradesh state in central India[1]. Instead of using live trapping to mark and release the tigers, the researchers used automatic cameras to "trap" tigers, and individuals were recognized in subsequent photos (Captures) by their unique pattern of stripes (marks). There were a total of 803 camera-trap days over a 3 month period (see Chapter 9). During this period, 26 individual tigers were identified. Several tigers were "captured" on photos 2 or 3 times, others only once, allowing the researchers to generate a capture history for each tiger (Appendix B). Because of the short study period (3 months) relative to the lifespan and gestation period of tigers, it is reasonable to assume the population was closed.

- Run CAPTURE with the data from Appendix B (follow the directions for exercise 1).

- CAPTURE output starts with a summary of the input data (check this to be sure it is accurate).

- CAPTURE carries out model tests. Look them over, remembering that a large probability value means that the data fit the model well. The first three tests all compare M_o against other models. M_o compares well with the other models (probabilities > 0.1). The remaining tests have probabilities around 0.05 and do not provide support for the other models.

- CAPTURE combines the results of the goodness-of-fit tests and uses an algorithm developed from simulated data sets to calculate the most appropriate model. Recall that the most appropriate model has a score of 1. The model tests for the tiger data are summarized below:

Model	M_o	M_h	M_b	M_{bh}	M_t	M_{th}	M_{tb}	M_{tbh}
Criteria	1.00	0.87	0.22	0.52	0.00	0.27	0.340	0.64

QUESTIONS:

- What is the most appropriate model?

- What is the estimated probability of capture, p-hat for the most appropriate model?

- What are the population size estimates and approximate 95 percent confidence intervals for this model?

- What model would you choose that would be more robust than the one considered in question 1?

- What are the values for population size and 95% confidence intervals for the second best model?

- The M_o model assumes that the capture probability for all tigers are equal, is that a reasonable assumption? Note: this would mean that all tigers behave the same way.

CAPTURE also provides several alternative M_h models, the most robust being the Jackknife M_h model. Locate the output for the Jackknife M_h model and find:

Average p-hat

Interpolated population estimate with standard error

Approximate 95 percent confidence interval

- How does the Jackknife M_h model values differ from the M_o model?

· Which might be more realistic.

· What is your best estimate of the number of tigers in Kanha National Park, with an approximate 95% confidence interval.

· Kanha National Park is 282 km[2] in area. What is the density of tigers in the park given as tigers per 100 km[2] ?

[1]modified from http://www.wcsmalaysia.org/analysis/Mark-recap_tigers.htm.

OPEN POPULATION MODELS

USING THE PROGRAM JOLLY

The Jolly-Seber model also uses capture-recapture data, except you are assuming the population is open. In fact, the program computes parameters such as survival rates and births for the interval between sampling sessions. As in the examples using DISTANCE above, a goodness-of-fit test is used to determine if this model is appropriate. In addition, JOLLY provides a model (Model A) which allows for death but no immigration, a model (Model 2) that assumes survival rates are different for animals captured for the first time as opposed to unmarked or previously marked animals, a model (Model B) that assumes constant survival rate and time-specific capture probability, and Model D, which assumes survival rate and capture probability are both constant.

In JOLLY, there are the following variables:

N_i = Population size at time *i*

SEN_i = Standard error of the population size estimate

PHI_i = Probability that an animal survives from time *i* to *i+1* (between two sessions)

$SEPHI_i$ = Standard error of the survival rate estimate

$COVPHI_i$ = Covariance between survival estimates for times *i-1, i*

M_i = Number of marked animals alive at time *i*

p_i = Probability that an animal alive at time *i* is captured in the *i-th* sample

B_j = Number of new animals added during the interval *i* to *i+1*, and alive at time *i+1* (e.g. by birth or immigration)

SEB_i = Standard error of the recruitment estimate

Data input:

The input file for program JOLLY contains both commands and data records. The control records are used to tell the program the title of the data, number of capture periods, interval lengths and data type. The format is:

```
TITLE=Your title here
PERIODS=6
FORMAT=(6I1,I4)
END
```

The data may be in one of three forms, but you will use the default - capture history record format. Capture history records are strings of zero's, one's and two's to indicate:

0 = animal not captured

1 = animal captured and released

2 = animal was captured but not released (e.g. animal was dead)

The data are similar to those used in DISTANCE but you now have a number that follow the data string. For example, the third line of data in your example is 110000 21. This is read as 21 animals have a capture history where they were captured in the first and second sessions, but not after that. Files must end with the END command.

EXERCISE 1: *MICROTUS* MARK-RECAPTURE

1) Go to PROGRAM JOLLY on the Patuxent Software page at

http://www.mbr-pwrc.usgs.gov/software/jolly.html.

2) Open and type in (or copy and paste) the contents of Appendix C in this chapter into the white data entry box and click the *Preform Analysis* button.

3) Save and print your file now to avoid losing it.

Let's look over the output. This output begins with a list of the data records that were input, followed by a few summary lines. The next part is the B-table summary. It should look familiar if you completed Chapter 6. Familiarize yourself with the contents of this table. It should look like Table 7.2 below.

The next few pages of the output present the results of the models computed by JOLLY. The first is Model A, followed by Model D, and the last is Model B. You should be able to look over these model outputs and locate the estimates of N, B, Phi, and p under each model. Compare each model and write a summary of your findings.

Table 7.2 Mark-recapture data from the program JOLLY for the Microtus data.

Time of last capture	Time or Recapture					
	1	2	3	4	5	6
1	0	44	1	0	0	0
2			33	4	0	1
3				32	1	0
4					28	4
5						38
6						0
Marked	0	44	34	36	29	43
Unmarked	56	28	15	21	17	34
Caught	56	72	49	57	46	77
Released	53	69	48	56	45	76

BIBLIOGRAPHY

Karanth, K. U., Nichols, J. D., Kumar, N. S., Link, W. A. and J. E. Hines. (2004) Tigers and their prey: Predicting carnivore densities from prey abundance. *Proceedings of the National Academy of Sciences* 101:4854-4858. (Available on line at http://www.ncbi.nlm.nih.gov/pmc/articles/PMC387338/)

Rexstad, E. and K. P. Burnham. (1991). *User's Guide for Interactive Program CAPTURE.* Colorado Cooperative Fish & Wildlife Research Unit, Colorado State University, Fort Collins, Colorado.

Williams, B. K., Nichols, J. D. and M. J. Conroy. (2001) *Analysis and Management of Animal Populations.* Academic Press, San Diego, CA.

White, G. C., Burnham, K. P., Otis, D. L. and D. R. Anderson. (1978) *User's Manual for Program CAPTURE.* Utah State Univ. Press, Logan, Utah.

APPENDIX A:

```
TASK READ CAPTURES OCCASIONS=6 X MATRIX
FORMAT='(A1, 1x, 6F1.0)'
READ INPUT DATA
a 100010
b 100000
c 100010
d 110001
e 111000
f 110100
g 101100
h 111110
i 101110
j 110111
k 101100
l 010100
m 010010
n 011001
o 011100
p 011111
q 001101
r 011011
s 010111
t 010100
u 010001
v 001000
w 010011
x 001001
y 001110
z 001111
task closure test
task model selection
task population estimate ALL
```

APPENDIX B:

```
TASK READ CAPTURES OCCASIONS=10 X MATRIX
FORMAT='(A1, 1x, 10F1.0)'
READ INPUT DATA
a 1001000110
b 1000000101
c 1101000011
d 0110000111
e 0101000001
f 0100000000
g 0010011000
h 0011000000
i 0001000011
j 0001100110
k 0001001001
l 0000100000
m 0001000000
n 1001000000
o 0001000000
p 0001100000
q 0001001000
r 0000110000
s 0000100000
t 0000010000
u 1000101000
v 0100000000
w 0001000001
x 0000010000
y 0000000001
z 0000100000
task closure test
task model selection
task population estimate ALL
```

APPENDIX C:

```
TITLE=Microtus
PERIODS=6
FORMAT=(6I1,I4)
END
100000  8
200000  3
110000  21
120000  2
111000  5
111100  3
111200  1
111110  1
111111  10
110111  1
101100  1
010000  10
020000  1
011000  3
011100  3
011111  3
011101  3
011011  1
010111  2
010100  1
010001  1
001000  7
002000  1
001100  3
001111  3
001112  1
000100  13
000110  2
000120  1
000111  4
000101  1
000010  4
000011  13
000001  34
END
```

8

Transects - Distance Sampling Using DISTANCE

TIME REQUIRED

One lab period of 3-4 hours for conducting the transect field study and another 3-4 hour lab period for the in-lab data analysis using DISTANCE software.

LEVEL OF DIFFICULTY

Moderate

LEARNING OBJECTIVES

Learn how to set up a line transect study.

Understand how to record data for a line transect study.

Understand the difference between indirect and direct counts.

Use program DISTANCE to analyze both indirect and direct transect data.

EQUIPMENT REQUIRED

30m tape-measure or topofil (hip-chain)

Fluorescent vinyl flagging tape

Marker pens

Maps of the study sites

Handheld GPS

Altimeter and clinometer

Optical range finder or survey laser binoculars

Data sheets and clipboard

Field notebook

Camera (optional)

Access to computers connected to the Internet

BACKGROUND

Sampling mammal populations in large areas (e.g. national parks or reserves) is difficult because it is impossible to count every animal in the area. Researchers must use a subsample of the entire population. Using quadrants or grids is not feasible for large mobile mammals, so researchers typically use line transects for these types of surveys. Transect lines are established in the area of interest, and each line is methodically and uniformly searched for the target species. If you could be assured of counting every animal along a transect of fixed width, then the line transect is simply a very thin rectangular quadrat (Krebs, 1999). However, it is unlikely that the observer will be able to count every animal along a transect (e.g. undetected animals, see Buckland et al., 1993). In such cases, we need to know the probability of detection before we can estimate density.

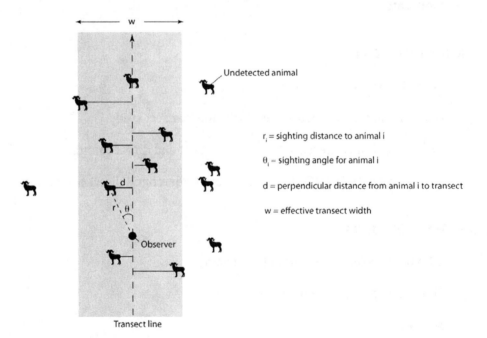

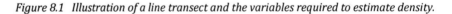

Figure 8.1 Illustration of a line transect and the variables required to estimate density.

When an animal is observed, three measures are taken:

r_i = is the sighting distance, from the observer to animal *i*

θ_i = sighting angle, between the animal, the observer, and the transect line.

d_i = perpendicular distance from animal *i* to the transect line

(calculated as d = r sin θ_i)

Notice that in Figure 8.1 not all individuals are detected as the observer moves along the transect. In fact, the probability of detecting an object decreases as its distance from the line increases. Thus, we need a way of estimating the detection function in order to have an accurate measure of the final density of animals. The detection function, g(x) = probability of observing an animal at "x" distance from the transect line.

We then estimate density (D) as:

$$\hat{D} = \frac{n}{2La}$$

where,

n = number of animals observed

L = total length of transect

a = 1/2 the effective transect width (e.g. the area under the detection function)

The basic problem in estimating density is to estimate the parameter a, and there are numerous methods for estimating a in the literature (see Krebs, 1999). In this exercise we will use software called DISTANCE to estimate a and provide an estimate of the total density of animals.

INDIRECT DATA

In some cases it is not practical to survey animals by direct observation. For example, some species are secretive and rarely seen, while others range over vast areas. In these cases it may be possible to use indirect evidence of the animals presence to estimate density from transect surveys. Typically, this involves using signs such as feces (pellet piles of deer or dung piles of elephants) or nests (of chimpanzee and orangutans). Such indirect methods require knowing 1) the defecation rate (or nest building rate), the number of dung piles produced per animal per day, and 2) the decay rate (e.g. how long it takes for the dung to disappear in the environment).

Dung count estimates have several limitations (Plumptre, 2000). First, it is often difficult to determine the species responsible for the dung pile. Second, defecation rates vary with diet. Third, weather conditions (and dung beetle activity) greatly affect the decay rate of dung piles. Finally, dung piles are often used to mark territories, and are not randomly distributed (leading to statistical problems). Nevertheless, dung counts may be the best option for large mammal surveys in forested areas. Dung counts have been used for surveying forest elephants (Barnes and Jensen, 1987), buffalo and forest duikers. For elephants, dung count density estimates appear to correlate well with those from other methods (Barnes, 2001). In part II of this exercise, we will use data from indirect counts of elephant dung to estimate population size for Asian elephants.

FIELD PROCEDURES

To survey large areas quickly or with few personnel (two or three surveyors) there are two options; a reconnaissance survey or line transects. Reconnaissance surveys are quick-and-dirty assessment that do not provide sufficient data for estimating density, but may be a first step to further census of the area. Unlike transects, reconnaissance surveys follow a winding path of "least resistance" through the habitat to cover as much ground as possible. Line transects provide data for estimating relative abundance or population density. Transect data can be collected for direct observation of animals or indirect observation via dung or other sign (see exercises below).

There are a few general guidelines for conducting most line transect surveys (see Anderson et al. 1979):

- Surveys should be carried out by night and day if nocturnal species are present.

- New transects should be located to sample different vegetation types and/or levels of disturbance.

- Transect(s) should be straight and well marked (a series of straight line segments may be more appropriate for some areas).

- All animals directly on the transect line must be seen (e.g. with probability equal to one).

- All distances and angles must be accurately measured.

- At least 40 animals should be seen (n > 40), and 60-80 are preferred. Therefore, transect surveys can only generate population estimates for species that are relatively abundant and visible.

- A pilot survey is recommended to aid in planning the survey design. Often, a simple visit to the area to be surveyed, along with basic biological information about the animal and its habits and habitat, will be sufficient to design an adequate survey to estimate density.

- Avoid transects running along roads, streams, or other potentially biasing features.

EXERCISES

EXERCISE 1: CONDUCTING TRANSECT SURVEYS

Transects should cover the main habitats in the approximate proportions as they occur in the area and should not cross each other. Transects should be positioned at least 300–500m apart. New paths should be cut before surveying to avoid frightening animals during the survey and to avoid the use of animal trails, logging roads, etc. (which would bias the survey). Ideally, transects are cut by first selecting a starting point and direction using a random number table. However, if the goal is to sample all habitats in the area, this approach may not be possible. Each transect should be kept straight using a compass or GPS device to maintain a constant heading. Record and map any streams, buildings, roads, as you go (using a GPS unit). It is good practice to mark the distance along the transect every 50m by writing the distance on flagging attached to a tree along the transect line. (remove all flagging at the end of the survey period if the transect won't be re-used).

Daytime surveys should be conducted when animals are active, from just after dawn to about 11:00 am. If this is not practical, afternoon surveys can be carried out between 15:00 and 17:30 hours. To avoid bias, do not conduct surveys during the rain. Transect surveyors should walk slowly, scanning from side to side for movements and listening quietly for sounds. An average speed of 1 km per hour, with a pause every 100m or so is common. This does not include the time it will take to collect data if an animal is sighted. Before beginning each transect walk, record all of the information in the top part of the transect survey forms (Appendix). When an animal is sighted, record the time, species identity, group size, group spread, and sighting location along the transect, onto the data sheet. The opportunity to record subsidiary information such as activity, diet, height, age and sex of animals sighted, mixed-species associations, and vegetation features are also important and should not be wasted. Sighting distances (SD) and angles are then recorded using a pre-calibrated range finder and a sighting compass. Convert these to perpendicular distances r (sin θ_i) and record this on the data sheet. (Note: It may be easier to memorize the exact location of the animal, and walk to the point perpendicular to the transect and measure from this location). Finally, record any other observations, including the number of males and females, and juveniles in the group, or any behaviors on the data sheets (Appendix).

EXERCISE 2: DATA ANALYSIS USING DISTANCE

In distance sampling, you expect to observe only a subset of all animals present in the area. The probability of detection for animals on the transect line itself is assumed to be 1, and the probability of detection decreases the farther the animal is from the transect line (Figure 8.2). The perpendicular distance from the transect to each animal is used to fit a model for the detection probability to the data. The model allows you to calculate the effective strip width (a), which is used to estimate density using the equation above.

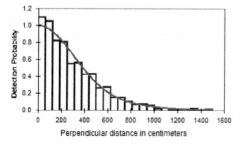

Figure 8.2   A graphical model of the detection function (curve) for a set of perpendicular distance from transect (histogram bars).

Estimating detection probabilities involves complex mathematical modeling. You will use a computer program called DISTANCE to calculate these models and provide a density estimate. DISTANCE can be downloaded from the Internet at:

(http://www.ruwpa.st-and.ac.uk/distance/distance60download.html).

DISTANCE runs under the Windows operating system. Mac users can run it using Parallels 5.

Transect survey data should be stored in a spreadsheet or database, and then imported into DISTANCE. The data for this example are stored in a *Microsoft Excel* file. DISTANCE cannot import directly from *Excel*, so you have to first export the data to a tab delineated text file.

Use *Excel* to open the file Example1.xls.

This file is located in the Sample Projects folder (usually "C:\Program Files\ Distance 6\Sample projects").

Note the layout of the file: There is one column each for stratum name, stratum area, transect name, transect length, perpendicular distance, and cluster size. (e.g., cluster size would not be required if objects were individual animals.) See Figure 8.3.

◇	A	B	C	D	E	F
1	Stratum	Area	Transect	Length	Distance	Cluster size
2	Stratum 1	100	Line 1	5	7.8595	1
3	Stratum 1	100	Line 1	5	10.2435	4
4	Stratum 1	100	Line 1	5	12.4435	2
5	Stratum 1	100	Line 1	5	3.76	3
6	Stratum 1	100	Line 1	5	4.7777	7
7	Stratum 1	100	Line 1	5	8.4531	3
8	Stratum 1	100	Line 1	5	13.4136	2
9	Stratum 1	100	Line 1	5	5.8039	2
10	Stratum 1	100	Line 1	5	7.4507	5
11	Stratum 1	100	Line 1	5	11.4513	3
12	Stratum 1	100	Line 1	5	0.8634	4
13	Stratum 1	100	Line 1	5	9.2329	6
14	Stratum 1	100	Line 1	5	12.5082	4
15	Stratum 1	100	Line 1	5	6.0755	8
16	Stratum 1	100	Line 2	2	9.1468	1
17	Stratum 1	100	Line 2	2	6.3828	2
18	Stratum 1	100	Line 2	2	21.2129	3
19	Stratum 1	100	Line 3	6	3.8497	4
20	Stratum 1	100	Line 3	6	12.5534	4
21	Stratum 1	100	Line 3	6	4.698	2
22	Stratum 1	100	Line 3	6	17.9612	3

Figure 8.3   The sample data file "example1.txt".

There is one row for each observation. A few rows have missing data (e.g. row 102), representing a transect where there were no observations.

· In *Excel*, choose *File | Save As...* Under Save as type: choose *Text (Tab delimited) (*.txt)*.

· Click *Save*, and now close *Excel*.

· You should now have a text file Example1.txt in the same folder as Example1. xls. (It is important that the new file be in the correct location).

· You can open the text file in a text editor (e.g., Notepad) to examine it if you like.

· Open DISTANCE 6.0. In DISTANCE, survey data and analyses are stored in a project. Before you can import the data, you first need to create a new project.

· In Distance, choose *File | New Project*. A window opens asking for the name of the project to create. Under File name, type "*Example1*", and click *Create*.

· The New Project Setup Wizard now starts (Figure 8.4). This is designed to guide you through creating a new project. The first step is to check the box next to *analyze a survey that has been completed*, click *Next* at the bottom of the window to continue. Step 2 gives some introductory information. Once you have read it click *Next*.

· Step 3 asks about the Survey Methods. Fill it in as shown in Figure 8.5. Make sure *Line transect* is selected. The observer configuration should be *Single observer*. Measurement type for this example is *Perpendicular distance*, and the observations are *Clusters of objects*. When the correct options are selected, click *Next*.

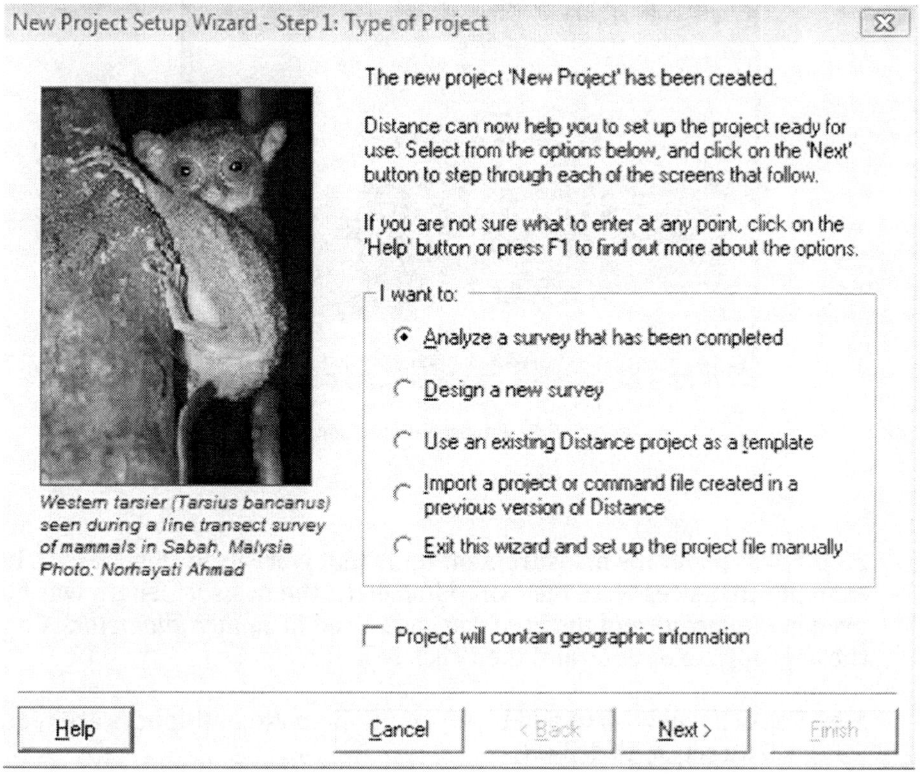

Figure 8.4   *Project Setup Wizard window Step 1.*

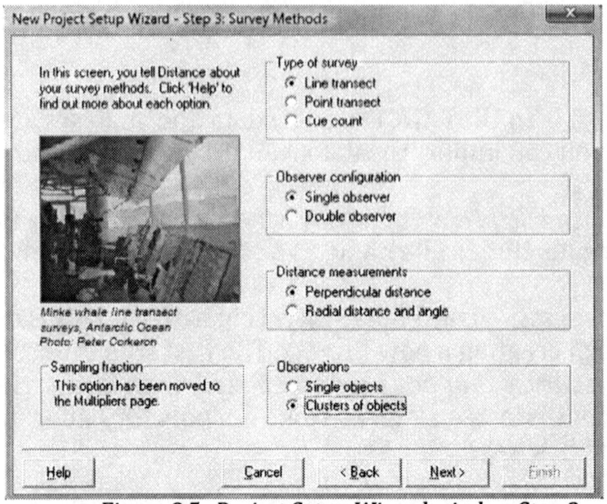

Figure 8.5   Project Setup Wizard window Step 3.

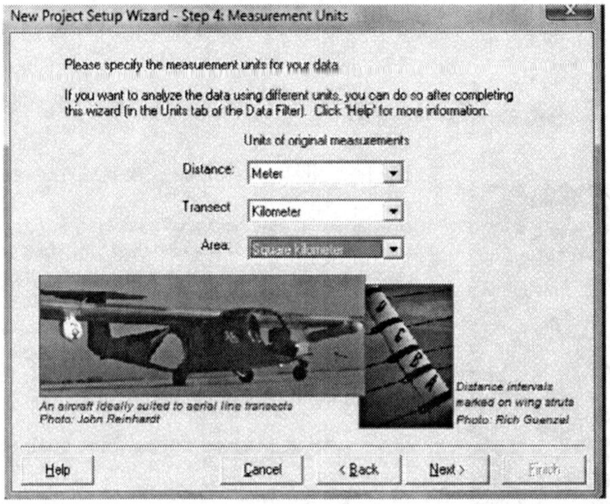

Figure 8.6   Project Setup Wizard Step 4.

· Step 4 asks about the measurement units that were used (Figure 8.6). In this example, distances were measured in *Meters*, the transect length was measured in *Kilometers* and the area was measured in *Square kilometers*. Choose the appropriate options and then click *Next*.

· Step 5 asks if you wish to add in any multipliers. No multipliers are required for this data set, so click *Next*.

· DISTANCE is now ready to set up the project, and in the last window (Step 6) asks you where to go next (Figure 8.7). You want to import some data, so select the option to *Proceed to Data Import Wizard*, and click *Finish*. The project is now created and the Import Data Wizard is started.

· If you have followed the previous steps, the Import Data Wizard is now open at the Introduction page (Step 1). Once you have read the text there, click *Next*.

· A window will open prompting you for the file to import. Find and select your previously saved file *Example1.txt* by clicking on it and then click *OK*.

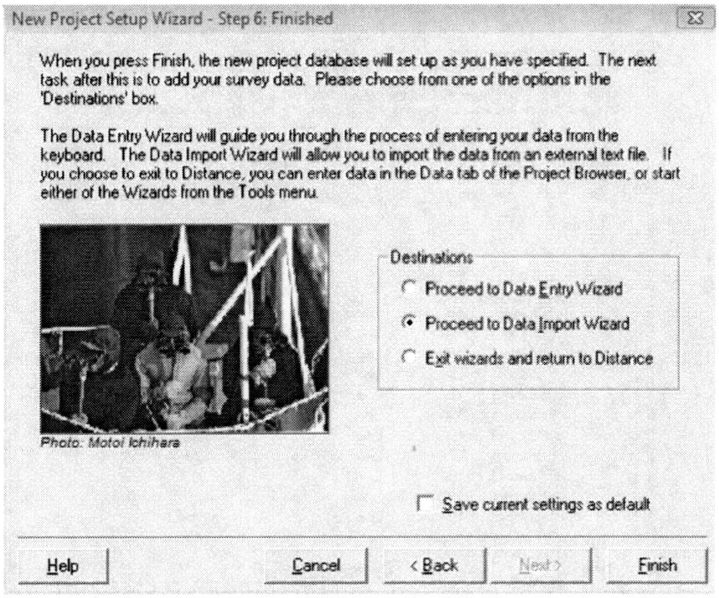

Figure 8.7   Project Setup Wizard Step 6.

· The Data Destination page (Step 3; Figure 8.8) of the wizard will open. This page asks you where the data from your text file should be placed in the Distance database. The default options here are appropriate for this example, so click *Next*.

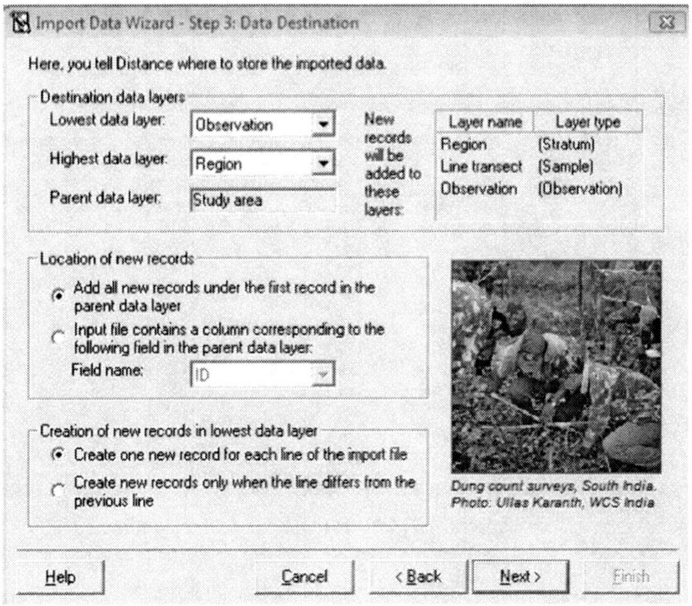

Figure 8.8   The data destination window.

· The Data File Format page (Step 4) is now opened (Figure 8.9. In our case, the first row contains the column labels, so click on the option *Do not import first row*, then click *Next*.

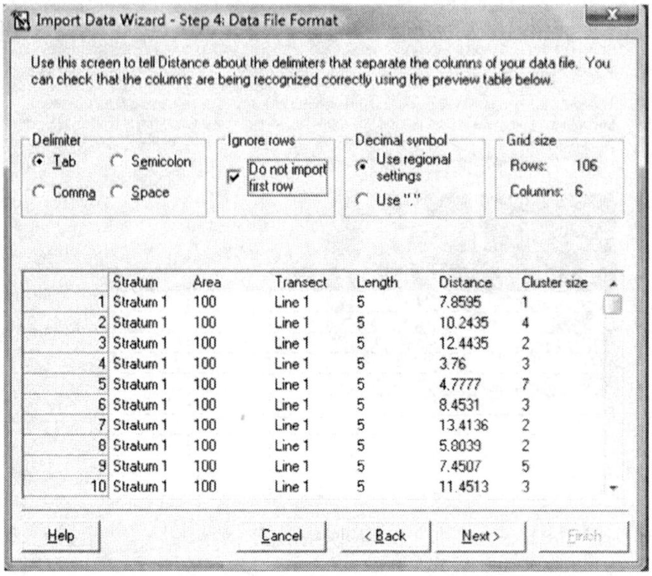

Figure 8.9   The data file format window.

· The Data File Structure page (Step 5) will now open up (Figure 8.10). Here you have several choice as to how you wish to proceed.

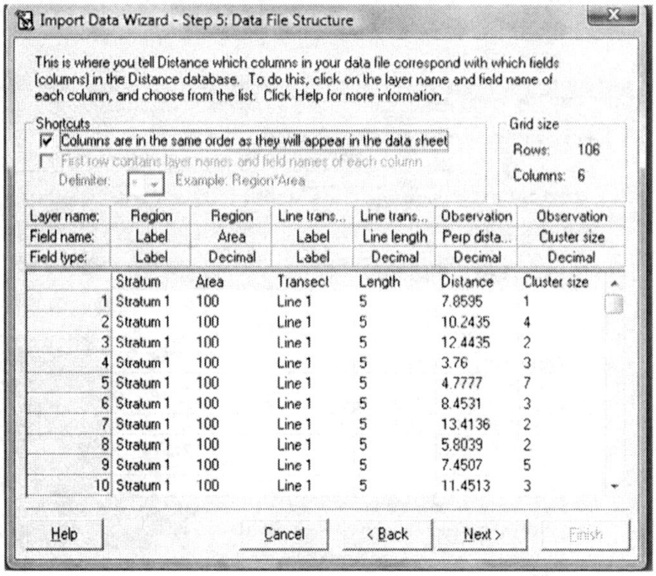

Figure 8.10   The date file structure window.

· If, like in this example, your data is set up in columns in the same order as you wish them to appear in the Distance data sheet checking the option *Columns are in the same order as they will appear in the data sheet* means that DIS-TANCE will automatically assign a layer name, field name and field type to each row.

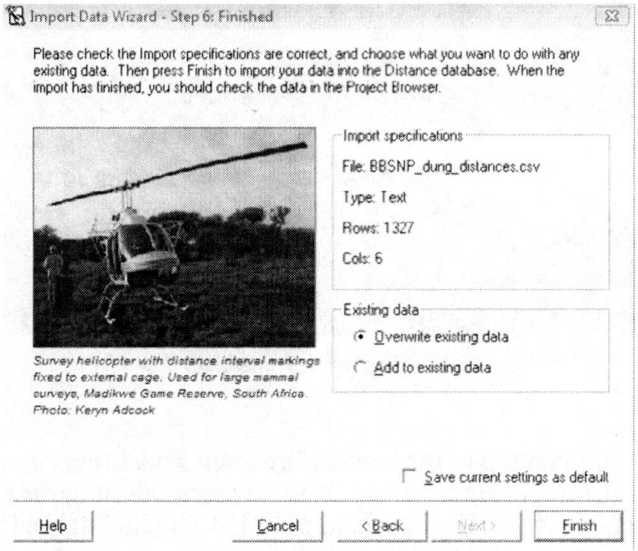

Figure 8.11   The final window in the import data wizard.

· Now click *Next*, and then click *Finish* to import the data (Figure 8.11). Now the data has hopefully imported successfully, and the project you created should be open with the Data tab of the Project Browser selected (Figure 8.12).

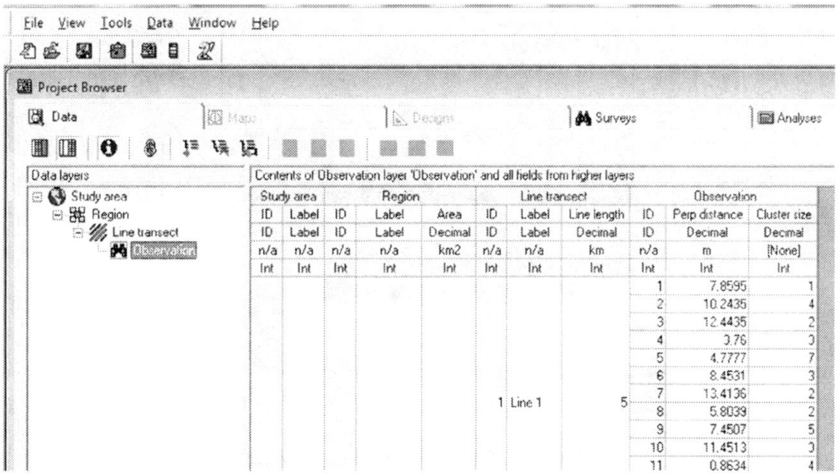

Figure 8.12   The project browser with the data entered.

· This is the main interface to your data in DISTANCE, and is called the Data Explorer.

· In the left-hand pane, under Data layers, click on *Observation* (Figure 8.12). This expands the view in the right-hand pane to show you all the data (Figure 8.13). Scroll down to verify that all 12 transect lines and 105 observations have been imported. Note that line 11 has no observations associated with it, as we wanted. Now it's time to do some analyses.

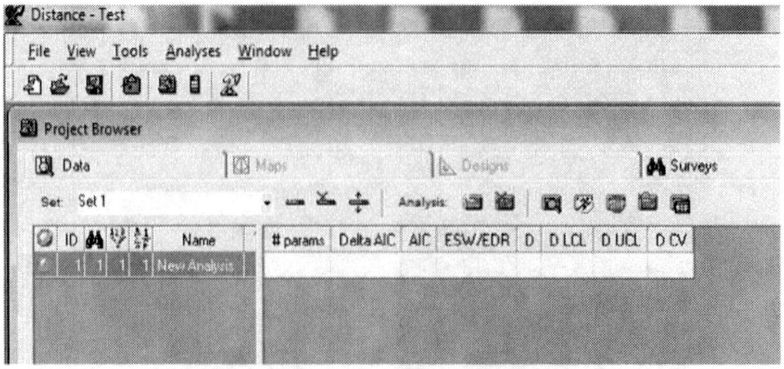

Figure 8.13   The project browser window.

· Click on the *Analyses* tab of the Project Browser. This brings up a table called the Analysis Browser (Figure 8.14). This is where the analyses that are carried out will be listed. In the left-hand pane is a line highlighted that says New Analysis. This is the default analysis that Distance creates. A grey ball (the status ball) also on that line indicates that the analysis has yet to be run.

· Double click on the *grey ball* (or choose *Analyses | Analysis Details*). A new window titled Analysis 1 will open up (Figure 8.14). This window is the Analysis Details window, and you are on the Inputs page (see tabs along the side).

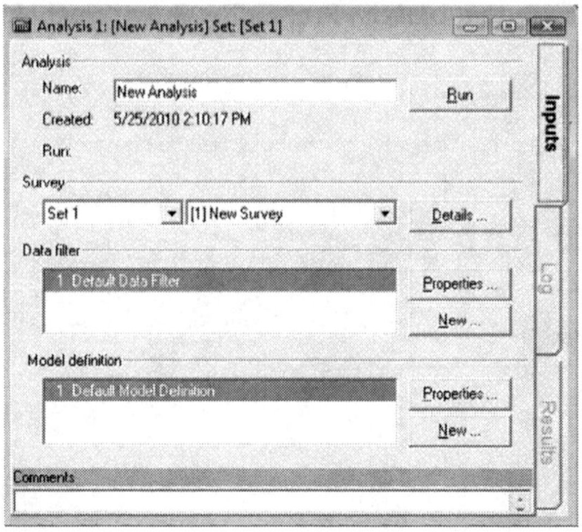

Figure 8.14 Analysis Browser window.

· At the bottom of the Inputs page is a section called Model Definition. The selected model definition is called Default Model Definition and this is the only one that has been defined so far. Click on *Properties* beside the model definition to open up the Model Definition Properties window. This window gives all the options available to change a model in DISTANCE.

· Click on the *Detection Function* tab. This will show what key function and adjustment term are currently selected for use. Explore the other options if you wish, then click *OK* to exit the Model Definition Properties.

· Click on *Run* in the Analysis Details Inputs page. Once the analysis has run the Results tab will turn green indicating that there were no problems with the analysis, and you will be taken to the Results page (Figure 8.15). If there had been a problem, then you would be taken to the Log page of the Analysis Details window. Then the log tab would be colored either amber for a warning, or red for an error.

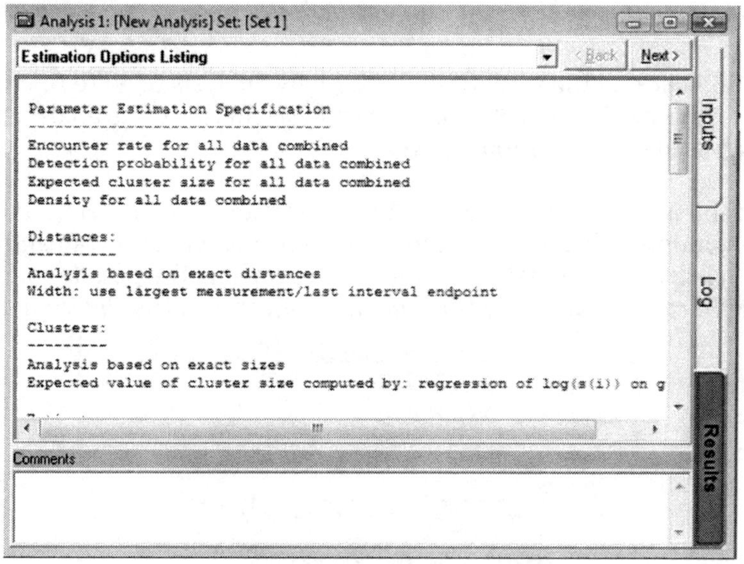

Figure 8.15 The results details window.

· In the Results tab click *Next* several times to view each page results. Once you have finished looking at the results, close the Analysis Browser window by clicking on the *Close* button in the top right hand corner of the window, or choosing *Analysis Results | Close*.

· In the Analyses Browser, the grey ball should now have turned green, to indicate that the corresponding analysis ran properly. A summary of the results is given in the right hand window pane. This is useful for when you wish to compare various analyses, without having to scan through all the results pages. (The columns shown can be configured by choosing *Analyses | Arrange columns. . . .*)

CHANGING MODELS IN DISTANCE:

The default analysis you ran in the last section uses a half-normal key function with cosine adjustments. Here, you'll create another analysis that uses a hazard rate key function with simple polynomial adjustments. Don't worry about the fancy terminology for now.

Return to the Analysis Browser (Analysis tab of the Project Browser; Figure 8.13) and click on the *New Analysis* button on the Analysis Browser tool bar (or choose *Analyses | New Analysis. . .*).

· Before you go any further, give the new analysis a sensible name that reflects what it does by clicking on the current name ("*New Analysis 1*") and then typing in the new name (e.g., "Untruncated hr+poly"). You might want to give a sensible name to the default analysis too (e.g., "Untruncated hn+cos"). This might not seem important when you only have two analyses, but if you don't name new analyses as you create them you'll soon find yourself loosing track of which is which!

· Double-click on the status ball for the new analysis to go to the Analysis Details window (Figure 8.14) for this analysis. Because the analysis is not run, you are taken to the Inputs page.

· Click on *New. . .* in the Model Definition section. The Model Definition Properties window will open up. Choose the *Detection Function* tab and under "Models", click on the model listed and change it using the pull downs to choose *Hazard rate key* and *Simple polynomial adjustments* as shown in Figure 8.16.

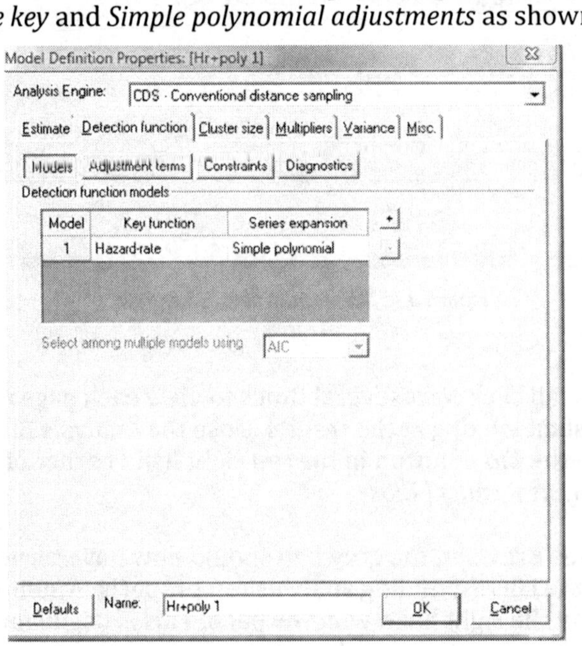

Figure 8.16   The model definition properties window.

· This new model definition is currently called "Default Model Definition 1", so give it a sensible name e.g., "hr+poly" by clicking on *Name*: at the bottom of the Model Definition Properties. Then click *OK*.

· Now run the analysis and compare the results with the previous one. You'll notice that the analysis shows a problem, as indicated by the orange tab for Results. The data need to be truncated. To truncate the data a new Data Filter needs to be created.

· In the Analysis Browser, click on the *New Analysis* button to create a new analysis.

· Name this analysis something like "19m trunc hn+cos".

- Double click on the status ball to open the Analysis Details for this analysis.

- Under Data Filter, click *New...*

- In the Truncation tab, select *Discard all observations beyond* and type in *19*. See Figure 8. 17.

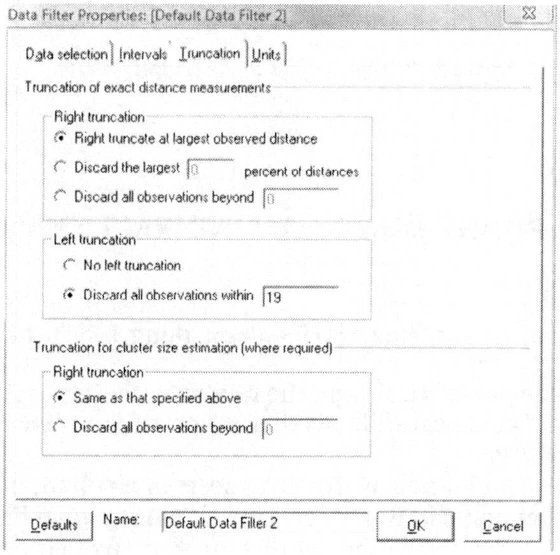

Figure 8.17 The truncation window.

- Give the new data filter a sensible name (e.g., "19m truncation"), and then click *OK*.

- Check that the highlighted Model Definition for this analysis is the "hn+cos" one.

Note that two analyses are now using the same Model Definition – this analysis and the first analysis we ran, which was without truncation. Let's not run this analysis yet – instead we'll create a second analysis with 19m truncation and hazard-rate + simple polynomial model.

- As you add analyses to the Analysis window they are stacked one on top of the other (Figure 8.18). Each displays the values the model calculated for:

- ESW/EDR is the effective strip width

- D is the estimate of the density of dung piles

- D LCL and D UCL are the lower and upper 95% confidence limits of density, respectively.

- D CV is the coefficient of variation for density.

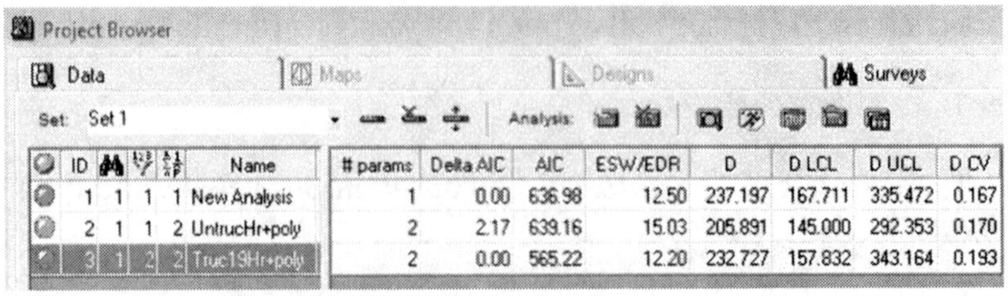

Figure 8.18   The analyses in the project browser.

EXERCISE 2: ELEPHANT DUNG - AN INDIRECT EXAMPLE

To begin, go to:

http://www.wcsmalaysia.org/analysis/Elephant_dung_DISTANCE.htm

and download the "lab guide" (pdf) and the raw data file (cvs format). Follow the instructions in the lab guide carefully, as this is a more complete example and has some interesting features.

The data set you will model is from transects of elephant dung (indirect method) for forest elephants in Bukit Barisan Selatan National Park in Sumatra, Indonesia. In this case, the perpendicular distances are from the transect line to elephant dung piles. DISTANCE is used to estimate detection probabilities, which are combined with encounter rates, dung defecation rates, and dung decay rates to estimate the density of elephants in the park (with confidence intervals). In 2001, Wildlife Conservation Society Indonesia Program estimated the elephant population of Bukit Barisan Selatan National Park (BBSNP) in Sumatra, Indonesia, from dung surveys (Hedges et al, 2005). The defecation rate (dung production rate) was estimated by observing 12 captive (but free-ranging) elephants in the nearby Way Kambas National Park. Dung decay rates (disappearance time) were based on monitoring 1,302 dung piles in BBSNP for 18 months prior to the transect survey. Finally, dung pile density in BBSNP was estimated by line transect surveys.

If each elephant produces p dung piles per day and dung piles remain visible for t days, then dung density will be:

$$S = D \times p \times t$$

where D is the density of elephants. If you can estimate p and t, you can calculate D from S (dung density). The preliminary surveys revealed that the defecation rate was 18.15 dung piles per 24 h with SE 2.53. Dung piles remained visible for 305.36 days with SE 7.33.

To sum up: you estimate the elephant population in BBSNP at 495 animals, with 95% CI of 331 to 738.

The population is concentrated in the "High" region, which makes up about 30% of the park, where the population is estimated at 440 animals (95% CI 290 to 667).

These figures are close to the results published by Hedges et al (2005): 498 animals, with 95% CI of 373 to 666.

To close DISTANCE, select '*File | Exit*' or click on the button in the top right corner of the main DISTANCE window. When the "Are you sure. . . " box appears, click '*Yes*'.

BIBLIOGRAPHY

Anderson, D. R., Laake, J. L., Crain, B. R., and K. P. Burnham. (1979) Guidelines for line transect sampling of biological populations. *Journal of Wildlife Management*, 43:70-78.

Barnes, R. F. W. (2001) How reliable are dung counts for estimating elephant numbers? *African Journal of Ecology*. 39:1-9.

Barnes, R. F. W. & K. l. Jensen. (1987) *How to count elephants in forests*. IUCN African Elephant and Rhino Specialist Group Technical Bulletin, 1, 1-6.

Buckland, S. T., Anderson, D. R., Burnham, K. P., and Laake, J. L. (1993) *Distance Sampling: Estimating Abundance of Biological Populations*. Chapman and Hall, London.

Hedges, S., Tyson, M. J., Sitompul, A. F., Kinnaird, M. F., and Gunaryadi, D. A. (2005) Distribution, status, and conservation needs of Asian elephants (*Elephas maximus*) in Lampung Province, Sumatra, Indonesia. *Biological Conservation*, 124:35-48.

Krebs, C. J. (1999) *Ecological Methodology*. 2nd edition, Benjamin/Cummings, Menlo Park, CA.

Plumptre, W. J. (2000) Monitoring mammal populations with line transect techniques in African forests. *Journal of Applied Ecology*, 37:356-368.

APPENDIX

Line Transect Data Sheet

Observer(s):		Date:	Field Sheet #:
		Site:	UTM:
		Latitude:	Altitude:
Vegetation:		Longitude:	Transect length
			Start time
Other			End time

Time	Distance along transect	Vegetation	Species	Animal Numbers			Sighting distance (r)	Sighting angle θ	Perpendic. distance (r*sinθ)	Observations:
				Males	Females	Juv.				

9

Camera Trapping

TIME REQUIRED

This exercise can be run over several weeks or months for the initial data collection phase. Data analysis can be completed in 3-4 hours.

LEVEL OF DIFFICULTY

Moderate

LEARNING OBJECTIVES

Learn how to use cameras as remote sensors in wildlife studies

Understand how to design and implement a camera trapping study

Gain experience placing cameras in the field

Learn how to collect and analyze camera trap image data

Use Internet-based software CAPTURE to analyze animal populations from camera trap data

EQUIPMENT REQUIRED

Camera traps (number depends on objectives and size of study area)

Extra batteries for cameras

SD data storage cards (minimum of 1 per camera)

SD card picture viewer (optional)

GPS device

BACKGROUND

Conservation biologist increasingly rely on rapid assessments of species richness and abundance, especially in regions threatened with habitat loss. These "quick-and-dirty" surveys are crucial for assigning conservation priorities. In the past these surveys have relied on track counts, short periods of live trapping, or direct observation of animals along line transects. While each method has its advantages, they all require considerable time and effort and do not work well for cryptic or wide-ranging species. Remotely triggered cameras (e.g. camera traps) have been used with great success in recent years to survey large tracts of remote areas. Networks of motion/heat-sensitive cameras are efficient and cost-effective way to collect data for species inventories, and are especially important for recording the presence of cryptic animals (Balme et al., 2009; Karanth and Nichols, 1998).

Camera trap networks are:

- Non-invasive

- Relatively simple to deploy

- Operate without supervision for weeks or months

- Provide permanent records of species, date, geo-referenced locations

- Video camera traps can also record animal behavior

- Cost effective over the long term

In addition to documenting presence of a species at a site (Figure 9.1), a properly designed camera trapping study, using carefully chosen, multi-year, geo-referenced locations can also document species distributions, local density (Rowcliffe et al., 2008), habitat preferences, and other population demographic parameters (Karanth et al., 2006). Therefore, camera-trapping is rapidly becoming a critical tool for wildlife conservationists (Karanth, 1995; Kays and Slauson, 2008).

Figure 9.1   Photo of a whitetailed deer taken with an inexpensive camera trap.

The most successful camera trap studies are those that are well-planned in advance. Researchers need to know what information they will need to answer their research or conservation question. In addition, it is vital to have knowledge of the geographic area and terrain to be sampled. Researchers should plan approximate camera positions and travel routes in advance. In addition, it helps to know the approximate home range size and basic ecology of the target species. Obviously, knowing how to properly operate the camera trap units and a GPS device is also vital. In most cases, a pilot study is required to become familiar with equipment, the logistics of moving through the terrain to check traps and replace batteries, and to get an estimate of capture success (number of photos/videos per unit time).

CAMERA SELECTION

There are now dozens of digital camera traps on the market. Choosing between the various models and types can be a daunting task. In the end, the choice will be dictated by the requirements of the study and the preference of the researcher. However, there are some basic considerations.

First, let's consider the type of camera. Digital cameras offer a number of advantages such as longer battery life, they can be used with infrared flash to capture nighttime images, and they store hundreds of images on tiny memory cards with no film development required. Not surprisingly, digital cameras have all but replaced film cameras in most studies (a drawback of less expensive digital cameras is that they can take a couple seconds to power up and take a photo after they have been triggered).

Digital camera traps can be triggered to capture an image using either active infrared (AIR) or passive infrared (PIR) triggers. Cameras using active triggering capture an image when an animal crosses a narrow infrared beam. In contrast, passive infrared triggers rely on the difference in temperature between an animal and it ambient surroundings. AIR camera traps are very effective at capturing target animals, but wind-blown leaves, flying birds, and rain storms can also trigger the cameras. PIR traps can also have false captures.

CONSIDERATIONS FOR CHOOSING CAMERA TRAPS INCLUDE:

Cost – Cost is always a consideration, but as with most things in life, you get what you pay for. In 2010, digital camera traps ranged from about $70 to over $500. The highest priced cameras come with lots of features and high megapixel ranges (higher megapixels means larger and shaper images). Before you buy the cheapest or most expensive model, carefully evaluate what your needs are. For example, you do not need a 5+ megapixel image to identify an animal. With typical camera trap surveys using 20-100 camera traps, cameras can quickly become a substantial budget item.

Trigger speed– Trigger speed is the time from when the infrared trigger first detects motion or temperature until the camera captures a photo. Rapid trigger speed is probably the most important consideration for large mammals. If it takes the camera 3 seconds to power up out of sleep mode and take a picture, the animal may already be gone (or you end up with a photo of its rump). Trigger speeds vary widely, with cheaper units having speeds as long as 6 seconds (useless for scientific studies). Trigger speeds of 1/2 second or less are best for wildlife surveys. If you must chose a camera with a slower trigger speed, make sure it has a wide detection zone so that the camera has a higher probability of taking a photo when the animal is still within the zone. Slower trigger speeds may still be useful at baited stations. For detailed reviews of trigger speeds see:

http://www.trailcampro.com/2009triggerspeedshowdown.aspx.

Detection zone– A camera trap's detection zone is the area within which movement is detected. Some cameras have narrow and long or short and wide detection zones (Figure 9.2). For most field studies a wide detection zone matched to the camera's lens field of view is best.

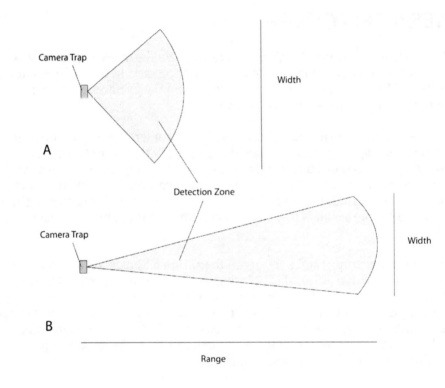

Figure 9.2   Comparison of detection zones for camera traps.

Recovery time– Recovery time is the minimum time between two consecutive photos (in a non-burst mode). Recovery times vary from a 1/2 second to 60 seconds. Researchers should consider if one photo every 60 seconds is adequate for their study design. For example, consider a scenario where a mother tiger triggers the camera and in the next few seconds she is followed by two cubs, who pass before the camera recovers. Such a scenario would result in the loss of important information for tiger conservationists. Recovery times of 1 second or less are generally acceptable in most cases. Rapid recovery cameras often capture multiple images of each animal which may be important if the animal carries visible tags or unique coat patterns used to identify individuals. (Remember, larger megapixel pictures require additional time to write to memory).

Flash range– Flash range is the maximum distance the flash projects to illuminate an animal with sufficient intensity to identify the animal. Manufacturers often report optimal values (up to 80 feet in some cases), but those are usually taken in ideal situations (e.g. open fields or in moonlight). The same camera located in a deciduous forest or on a cloudy night would lose as much as 50% of its advertised flash range. Infrared flashes are also available and less invasive to most mammals.

Battery life– Battery life is given (by manufacturers) as the number of days a camera is able to operate on a single set of batteries. There are many variables which affect battery life. It depends on the number of pictures taken per day, ambient temperature, camera settings, and many other factors. Consider using rechargable NiMH, NiCd, or Li-ion batteries for longer battery life.

Note: Most cameras take between 4 and 8 D or C cell batteries (test your batteries before placing them in the unit as many over the counter battery packs have at least one dead battery in the package).

OTHER FACTORS TO CONSIDER IN CHOOSING THE BEST CAMERA ARE:

Technical skill – Some camera traps have more features and require more expertise for proper field use. Consider who will be deploying the cameras and the level of training required.

Weather – Some camera traps are weatherproof units that can be submerged for short periods, but many units are only modestly water-resistant. Consider the environment and weather conditions carefully before purchasing camera traps.

VIDEO CAMERA TRAPS

Recent advances in video technology have reduced the size and cost of video devices to the point where many researchers are opting to use video camera traps. Many of the digital camera traps discussed previously also have the capacity to take short video clips and store those to memory. These may be pseudo-video if the number of frames per second is less than the standard 30 fps for video. In addition, the resolution of the video footage is often only 640 x 480 pixels per frame. For researchers wanting behavioral information, video capability may be a necessity. As always there are drawbacks, including shorter battery life, fewer events recorded to memory, longer trigger times, and greater expense.

SURVEY DESIGN

There are a number of criteria to consider in designing a camera trapping study. Foremost among them is meeting the studies objectives. Often there are trade offs involving the number of cameras available, the duration of the study, the coverage area, and statistical or other data analysis requirements. Deployment duration (days in use) and site coverage (size of the survey area) are primary factors in designing effective studies. The longer the deployment the greater the probability of detecting a given species, but longer deployment at one site also means that fewer sites can be surveyed in a given period of time (e.g. one season).

Camera traps have been used for a variety of research studies including to estimate density of target species, relative abundance of target species, relative abundance of mammal communities, species diversity, activity patterns, as well as to document the use of a specific habitat feature (e.g. burrow entrances or under-passes). Cameras can then be used to compare any of these metrics between different study areas (i.e. treatments), with a general rule of thumb of ~20 camera/sites per study area (at least for the more easy to detect species). In some circumstances, baited sets may be the preferred method of documenting the presence of a specific species.

Ideally, camera traps should be placed to uniformly cover the study area. The best approach is a hexagonal array because it maximizes the area covered with the fewest cameras (Figure 9.3). In many studies researchers are using 2 camera traps per site (pointed in different directions) to maximize capture probability at each site. How do we determine the approximate size of each hexagon?

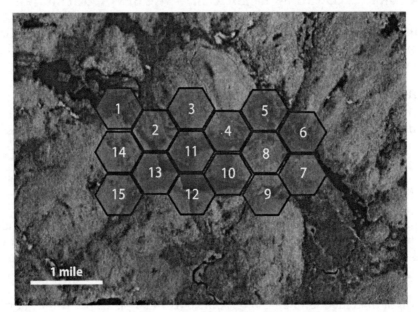

Figure 9.3 An example of a hexagonal array of 15 camera trap sites at a location in the Adirondack Mountains, NY.

The simplest answer is that camera stations (2 cameras each) should as far apart as possible, but with no gaps between camera stations larger than the target species minimum home range. This may be difficult in studies designed to survey mammals with very different home range sizes. For example, suppose a literature review suggests that the minimum home range of a leopard is 10 km² (ten square kilometers). This means that the maximum distance between camera sites can not exceed 3.6 kilometers; the diameter of a circle encompassing 10 km² is 3.6 km (Figure 9.4).

Of course, cameras can (and probably should) be placed closer together than this, but this is the maximum distance between adjacent cameras for an animal with such a home range. A survey concentrated in too small an area will capture few individuals and the sample size may be too small for statistical validity. Likewise, surveys spread out over too large an area for the number of camera traps will miss individuals (potentially violating assumptions of mark-recapture methods). In general, the survey areas should be large enough to capture many individuals (or species).

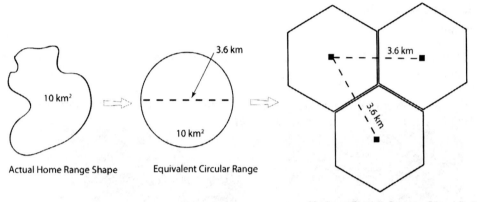

Figure 9.4   Calculation of the distance between cameras for a home range of 10 km².

The next decision is how long cameras should be deployed. This will depend on the remoteness of the site, logistics, and other considerations. If the goal is to estimate density using a closed population model (see Chapter 6), then cameras should be in the field for sufficient time to capture enough data without violating the model's assumptions. For example, mark-recapture models may require a closed population (i.e. no individuals can be added or removed by birth, death, or dispersal during the study). Long duration studies may violate this assumption. A second assumption of these models is that every individual has a nonzero probability of being captured (photographed). This is the main reason for spacing cameras in hexagonal arrays with inter-camera intervals less than the minimum home range area for the target species.

Note: This does no mean every individual must be photographed, only that each individual has the same chance of being photographed.

PLACEMENT AT CAMERA STATIONS

Before you go to the field, study a topographic map or satellite photos of the survey area. Locate and mark all trails, roads, waterholes, streams, etc. where there is a high likelihood of photographing your target species. Begin locating potential camera stations on the map by penciling in locations and spacing them appropriately.

Now look for gaps between cameras greater than your study design allows and move camera stations around to fill those gaps (e.g. you may need to move cameras due to lakes or other non-habitat within the study site). Once you have a potential map of you camera stations pinpointed on the satellite map, use *Google Earth* to record the exact coordinates (latitude/longitude) of each station. These "predetermined" coordinates will serve as a guide to finding the sites in the field using a hand-held GPS unit.

Remember:

· Cameras have to be monitored, so consider the logistical implications of your survey map.

· Camera placement need not be random; the final field position may differ slightly from those on your survey map.

Figure 9.5 A camera trap positioned on a tree along a deer trail.

Upon arrival at a predetermined camera station at your field site, look the area over carefully. Each location will offer unique challenges with respect to positioning the camera traps. Find the best possible location as close as possible to the predetermined coordinates. Look for sign (e.g. tracks, scat, etc) or game trails. Trails, dirt roads, river banks, waterholes, and game paths are all used regularly by many mammals. Choose sites where animals are likely to pass within range of the camera (Figure 9.5). Avoid open areas where the animal's path is difficult to predict. Keep in mind that humans also may follow these paths, and human presence may prevent animals form using these paths regularly. Humans may also remove or damage cameras. Spending even a few hours to find just the right spot is well worth the effort since each camera will be in the field for weeks or months.

Once you have found a suitable site for the camera station:

1) take a GPS reading and record the "actual" location of the station.

2) select one or two suitable trees to attach the units to. Trees should be relatively straight and thin enough to tie the traps in place (but not so thin as to move with the wind). Trees should be two meters from the most likely travel path.

3) make sure the ground is reasonably level and there are no obstructions in the field of view of the camera.

4) cameras should be positioned slightly down trail (rather than exactly perpendicular).

5) if using PIR traps, avoid sites where cameras will be in direct sunlight during sunrise or sunset.

6) consider the height of the animals and position the trap at about 20-30 cm from the ground.

7) inspect camera traps by checking to make sure that the date & time stamp is set correctly (photos are essentially useless without a date and time).

8) load the units with new (tested) batteries.

9) set the programmable features to the desired settings (e.g. image size, trigger delay, etc).

10) secure the camera traps to the tree using wire or chain. Attach a lock to the camera to prevent theft. For baited stations, aim the camera at the bait station and secure the bait so that the animal cannot easily remove it.

11) it is good practice to also attach a label that indicated the purpose of the study and provides contact information.

12) test cameras by walking in front of them (most camera traps have an indicator light that will light up when something is detected).

13) describe the station in your field notes.

EXERCISE 1: A CAMERA TRAP FIELD STUDY

Design a camera trapping protocol for a site selected by your instructor. Use the information provided here to determine camera placement and overall study design. Collect SD memory cards (digital photos) and replace batteries as needed on a regular schedule. Organize the photos into a database or spreadsheet file for analysis.

DATA ANALYSIS

Collecting Data: Photos are collected from each camera during routine monitoring of the study area. Photos can be collected from digital camera traps by:

- removing the SD memory card and replacing it with a new one. These used SD cards are labeled and returned to the lab and downloaded into a computer using a standard card reader.

- a computer is brought into the field and a USB cable attached to the camera trap to download the photos.

- a portable SD card viewer can be used to visually scan each SD card at the station and transfer the images to a second SD card.

Good record management is critical to a successful study. Each SD card should be labeled as to its station and camera number. Images from each camera should be entered into a database, keeping such information as camera number, station number, date, time, species, etc. as separate fields in the database.

Analyzing data: Camera trap data consist of date & time stamped images of a particular species at a geo-referenced location. Ideally, you would like to be able to identify individuals, but this is not always possible. For spotted and striped cats (e.g. jaguars, ocelots, leopards, tigers) the pattern of spots or stripes on the body can be used to identify individuals from photos. For more uniformly colored species (e.g. ungulates, coyotes, mountain lions, bears) is it generally not possible to identify individuals from photographic data. Nevertheless, camera trap data can be used to estimate density even without individual recognition (Rowcliffe, et al., 2008). These estimates, however, are complex and beyond the scope of this discussion.

One of the most basic measures is detection frequency (Figure 9.6), often defined as one photograph of a species per camera phototrap per day (24 hours). The date & time stamps on each photo also provide information of activity patterns (Figure 9.7).

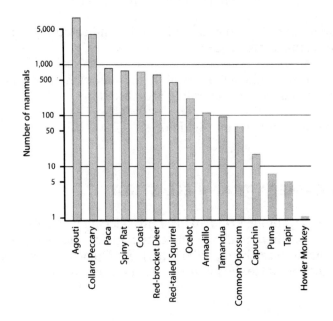

Figure 9.6   Frequency of detection for 15 tropical mammals on Barro Colorado Island, Panama. (adapted from Kays et al., 2009).

Another relatively simple measure is detection rate, the total number of photos of a species divided by the total time a camera was running (excludes periods of camera malfunctions). Summary information, including effort or trap-nights and species richness should also be tabulated for each study. An index of relative abundance index (RAI) can also be calculated from camera phototraps. The RAI for each species is calculated by summing all detections for each species for all camera traps over all days, multiplied by 100, and divided by the total number of camera phototrap days. For example, if 10 cameras are used for 100 days, then 10 * 100 = 1,000 phototrap days. If one black bear was photographed 4 times on day one and 1 time on day 4 and 1 time on day 5, then the RAI for the bear is calculated as:

$$RAI = \frac{(1+1+1)(100)}{10 \times 100} = \frac{300}{1,000} = 0.3$$

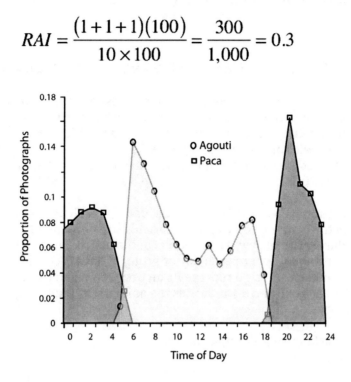

Figure 9.7   Activity patterns of two tropical rodents on Barro Colorado Island, Panama. Notice the diurnal and nocturnal patterns of each species. (adapted from Kays et al., 2009).

EXERCISE 2: DATA ANALYSIS FROM CAMERA TRAP STUDIES

Using the data you collected from your field site in exercise 1, calculate the species richness, detection frequency of each species, the detection rate, RAI, and activity patterns of species. Display your data graphically. Write a "Materials and Methods" and a "Results" section for a research paper that you might submit using this study design (follow the format for the *Journal of Mammalogy* - download the "information for contributors" document at

http://www.mammalsociety.org/pubjom/index.html).

If the target species carry unique coat markings (e.g. spot or stripe patterns) additional information can be gleaned from camera trap studies using mark-recapture methods (Karanth and Nichols, 1998). Mark-recapture methods and analysis is discussed in Chapter 6. With large data sets, it is more practical to use computer software programs to assist in the analysis of mark-recapture data. One of the most commonly used programs for analyzing abundance estimates using camera photographs is the program CAPTURE (Otis et al., 1978; White et al., 1982; Rexstad & Burnham, 1991). CAPTURE is available for download at

http://www.mbr-pwrc.usgs.gov/software.htmland

an online version is available at
http://www.mbr-pwrc.usgs.gov/software/capture.html.

EXERCISE 3: USING CAPTURE SOFTWARE FOR CAMERA TRAP DATA

You will use CAPTURE to generate abundance estimates based on mark-re-capture data from a hypothetical data set for a camera trap study. In this case each animal should be individually identifiable. The program CAPTURE uses several different models, each with different assumptions. The user's manual (at http://www.mbr-pwrc.usgs.gov/software.html) should be consulted to familiarize yourself with the models (see also Otis et al., 1978; Rextad and Burnham, 1991).

Data Preparation: CAPTURE requires the data for each animal to be in a matrix. Each row in the matrix denotes the capture history of each individual animal photographed during the survey. The columns in the matrix denote the sampling occasions for each animal, where each day or group of days is considered a sampling occasion. For each animal 0 represents an occasion when the animal was not captured (photographed), and a 1 indicates the animal was photographed on that sampling occasion.

Consider the following example matrix:

A 100
B 101
C 011

There are three animals each represented by a row, with the animal indicated by a letter (A-C). Animal A was photographed on the first day but not on day 2 or 3. Animal C was not photographed on day 1, but was photographed on days 2 and 3. CAPTURE also requires a format statement telling it how the data is formated and what models to run.

Here is an example of the formating headers for CAPTURE:

```
title='Hypothetical Ocelot Survey'
task read captures occasions=6 x matrix
matrix format='(a2,1x,6f1.0)'
read input data
FL 100110
BA 101010
EA 001000
OV 011001
RB 100010
VI 1001010
HA 010110
TA 101001
WB 010001
WO 011000
task closure test
task model selection
task population estimate ALL
task population estimate APPROPRIATE
```

What does all this mean? The first line is a title for the data set "Hypothetical Ocelot Survey". This is followed a task command, which says to read captures for 6 occasions in an X matrix. Matrix format is a description of how the data is formated in the matrix. For example, format='(a2,1x,6f1.0) refers to a two letter animal code (a2), a space of one character (1x) followed by a row of 6 data sessions in zero or one format (6f1.0). Read input data does just that - it reads the data that follows. The task statements that follow ask CAPTURE to preform some functions or apply a model to the data. For more detail go to this website:

http://www.mbr-pwrc.usgs.gov/software/capture.html.

The output of this run is 16 pages but includes the following:

- Number of trapping occasions was 6

- Number of animals captured, $M_{(t+1)}$, was 10

- Total number of captures, n, was 25

- Estimated probability of capture, p-hat = 0.4167

- Interpolated population estimate is 11 with standard error 1.2880

- Approximate 95 percent confidence interval 11 to 17

Try running the data set above by copying and pasting it into the data box in the online CAPTURE website at

http://www.mbr-pwrc.usgs.gov/software/capture.html.

Estimating density: CAPTURE generates an estimate of abundance, not density. To get density you have to take the abundance estimate (11 in our case above) and divide by the effective area sampled. It is common practice to add a buffer zone around the outside perimeter of the camera trap area. This takes into account the fact that some ocelots have home ranges primarily outside the camera trap area, but whose home ranges may partially overlap within the sampling area. There are many methods for adding buffer zones. One common method is to use a buffer zone width that is half the mean maximum distance moved (HMMDM) among multiple captures of individuals during the survey period (Karanth and Nichols, 2002; Karanth et al., 2004). Large regions that are not inhabited by the target species (e.g. bodies of water, villages etc.) should be subtracted to get an effective sample area (Figure 9.8).

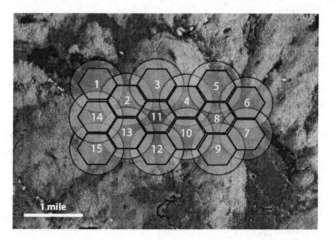

Figure 9.8 Camera trap grid with a buffer zone added (circles).

Note: CAPTURE generates standard error of the abundance estimate and a 95% confidence interval. These are measures of uncertainty associated with the abundance estimate. Abundance and density estimate for one study site should not be used to extrapolate densities to other study areas.

BIBLIOGRAPHY

Balme, G. A., Hunter, L. T. B. and R. Slotow. (2009) Evaluating methods for counting cryptic carnivores. *Journal of Wildlife Management.* 73:433-441.

Karanth, K.U. (1995) Estimating tiger (*Panthera tigris*) populations from camera-trap data using capture-recapture models. *Biological Conservation* 71, 333–338.

Karanth, K., and J. Nichols. (1998) Estimation of tiger densities in India using photographic captures and recaptures. *Ecology*, 79:2852-2862.

Karanth, K. U. and J. D. Nichols, (Eds). (2002). *Monitoring Tigers and Their Prey: a manual for researchers, managers and conservationists in tropical Asia*. Bangalore: Centre for Wildlife Studies.

Karanth, K. U., Chundawat, R. S., Nichols, J. D. and N. S. Kumar. (2004) Estimation of tiger densities in the tropical dry forests of Panna, Central India, using photographic capture-recapture sampling. *Animal Conservation.* 7:285-290.

Karanth, K., Nichols, J., Kumar, N. and J. Hines. (2006) Assessing tiger population dynamics using photographic capture-recapture sampling, *Ecology*, 87:.2925-2937.

Kays, R. W. and K. M. Slauson. (2006) Remote cameras. pp. 110-140, In *Noninvasive Survey Methods for Carnivores* (Long, R. A., MacKay, P., Zielinsky, W. J., and J. C. Ray, eds), Island Press, Washington, D.C.

Kays, R. W., Kranstauber, B., Jansen, P. A., Carbone, C., Rowcliffe, M., Foundtain, T. and S. Tilak. (2009) *Camera Traps as Sensor Networks for Monitoring Animal Communities.* The 34th IEEE Conference on Local Computer Networks (at http://www.nysm.nysed.gov/staff-pubs/docs/20118.pdf)

Otis, D.L., Burnham, K. P., White, G. C. and D. R. Anderson (1978). Statistical inference from capture data on closed animal populations. *Wildlife Monographs*, 62:1-135.

Rexstad, E. and K. P. Burnham. (1991) *Users Guide for Interactive Program CAPTURE*. Colorado Cooperative Fish & Wildlife Research Unit, Colorado State University, Fort Collins, CO.

Rowcliffe, J. M., Field, J., Turvey, S. T. and C. Carbone. (2008) Estimating animal density using camera traps without the need for individual recognition. *Journal of Applied Ecology*, 45:1228-1236.

White, G. C., Anderson, D. R., Burnham, K. P. and D. L. Otis. (1982) *Capture-recapture and Removal Methods for Sampling Closed Populations*. Los Alamos National Laboratory, Los Alamos, New Mexico, USA.

10

Radio Tracking

TIME REQUIRED

One lab period of 3-4 hours for collecting the radio tracking data and one lab period of 3-4 hours to analyze the data.

LEVEL OF DIFFICULTY

Moderate to High

LEARNING OBJECTIVES

Understand the principles of remotely tracking mammals.

Learn how to capture, immobilize, and collar small mammals for a radio tracking study.

Understand the principles involved in acquiring triangulation data.

Understand how to analyze radio tracking data.

Understand how radio tracking is used to study mammals in the field.

EQUIPMENT REQUIRED

VHF radio collars (minimum of one)

Receiver tuned to collar frequency

Yagi hand-held antenna tuned to collar frequency

Sighting compass

Two-way radios for communication between tracking stations

Live traps (optional)

Immobilizing drugs and equipment (optional)

Field note book (optional)

Permits and IACUC approval

BACKGROUND

There are three main types of telemetry used in wildlife tracking studies today: 1) radio tracking using VHF (very high frequency) signals; 2) satellite tracking using ARGOS or other satellite systems; and 3) global positioning systems (GPS) data loggers or transmitters. VHF wildlife tracking has been in use since the early 1960s and is the focus of this lab (Kenward, 1987; GPS tracking will be discussed in Chapter 11). Wildlife radio telemetry typically involves the transmission of radio signals sent from an animal carrying a radio transmitter to a researcher holding an antenna and radio receiver tuned to the frequency of the transmitter(s) (Figure 10.1).

Figure 10.1 A radio-collared bighorn sheep in Anza Borrego State Park, California. (Photo by Alan Vernon)

Radio telemetry can provide information on an animal's location, movement patterns, home range, migration patterns, habitat preferences, den or hibernacula sites, and in certain cases physiological parameters such as body temperature and dive depths (for marine mammals). Thus, there are a wide range of research questions that can be addressed using radio telemetry data. Nevertheless, before embarking on a telemetry study researchers should ask themselves:

· Is radio tracking is the best method available?

· How many animals must be tracked to acquire sufficient data?

· Can the species be readily captured (and recaptured)?

· Can the study species carry the transmitter without altering its behavior or survival?

· What size radio transmitter will be needed for the duration of the study?

· How often will subjects be tracked?

· How many people will be required to track the subjects of the study?

· Is the budget sufficient to cover the costs of transmitters, receivers, and personnel?

Among the most important of these questions are those involving sample size, both the number of animals to be tracked and the number of positions for each animal tracked. For example, one hundred tracking positions could come from one animal or from ten animals each with ten positions. The point is that researchers must consider the sample sizes before conducting the study to ensure that appropriate statistical tests can be used on the data collected. In general, it is the number of animals tracked that is most important (to avoid pseudoreplication), but there is obviously a compromise between tracking many individuals and collecting a sufficient number of positions for each individual. Thus, the objectives of the study must drive decisions about sample size. For example, studies on home range size and overlap may need to sample adequate numbers of both males and females.

Alternatively, studies that seek to understand migration patterns or seasonal use of habitats may need to track individuals over many months. It is important to remember that statistical tests often require that radio-positions be a random sample; this generally requires acquiring positions at random times throughout the study period or (more commonly) sampling at regular intervals. In certain cases, radio-tracking is used just to locate the individual or its group in order to conduct behavioral observations. In such cases, only one individual in the group needs to be tracked and the number of exact positions are less important (excellent discussions of telemetry can be found in Kenward, 1987).

TYPES OF RADIO-TELEMETRY STUDIES

HABITAT STUDIES

One of the most common uses of radio-telemetry is to understand habitat usage. The proportion of radio-positions in each habitat relative to the proportion of habitat available can indicate habitat preferences (or avoidance). Care should be taken to ensure that both sexes are radio-tagged, as habitat use often differs between the sexes. Likewise, it important to collect radio-positions during the full 24 hour day if the study involves species with nocturnal or crepuscular habitats, or habitats that differ between day and night. Habitat preferences may also vary dramatically between years. Ideally, a fixed schedule of sampling should be employed.

Each radio-position must also be accompanied by a habitat classification. Habitat can be classified in the field during the radio-tracking process, or later, when the researcher revisits each location. Alternatively, the positions can be plotted on a detailed habitat map of the region (if one exists). One important source of error in such studies concerns the size of the habitat patches and the size of the telemetry errors. Each radio-telemetry position is the result of triangulation and has an error polygon associated with it (Figure 10.2). If the habitat patches are smaller than the telemetry error polygon, then the proper assignment of habitat is difficult (Nams, 1989).

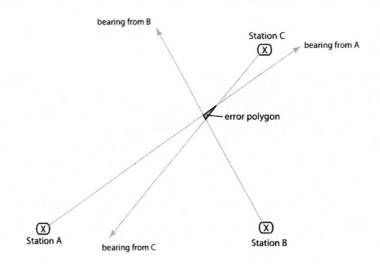

Figure 10.2 Error polygons associated with triangulation of radio-telemetry positions vary in size and must not be larger than habitat patch size.

LOCATING DENS, HIBERNACULA, OR OTHER FEATURES

Radio-tracking is an excellent way to locate dens, roosts, or feeding areas. Biologists, however, should take extra care when attaching transmitters to breeding individuals to ensure no disruption to the animal's survival and/or reproductive success. For example, pregnant females may be especially susceptible to the stresses associated with capture and transmitter attachment. In addition, following tagged animals to den sites may expose the litter to increased predation or observer disturbance.

HOME RANGE DETERMINATION

Another common use for radio-tracking data is to determine home range size for several representative animals in a population. As previously mentioned, it may be necessary (depending on the study objectives) to include different age/sex classes. If there is little variation in home range size among individuals of different age/sex categories, then it may be useful to pool the data from these groups. However, males often have significantly larger home range size than females, and in such cases it would not be appropriate to pool data from both sexes. Kenward (1987) suggested that approximately 30 relocation positions per individual is an adequate sample size for home range determination. In general, it is best to determine the appropriate sample size by conducting a pilot study or constructing an asymptotic plot of home range size versus number of positions. In practice, an animal's home range size and shape are determined by constructing a minimum convex polygon from the positions, or by using software for fitting polygons or kernel estimates (Seaman and Powell, 1996; Seaman, et al., 1998). Similarly, daily movements (distance and velocity) can be summarized from the distances between consecutive radio-positions.

DEMOGRAPHIC STUDIES

In theory, radio-telemetry data should help determine the cause of mortality, if the collared animal can be located soon after death. Survival rates can also be calculated from the number of transmitter-days (Pollock, et al., 1989). In practice, it is often difficult determine the cause of mortality or to distinguish mortality from tag loss or tag failure.

RADIO-TELEMETRY EQUIPMENT

TRANSMITTERS

Conventional radio-transmitters consist of transmitter unit, a battery (or solar cell), and an antenna. The specific components chosen for a study will depend of the study design and size of the study species. The transmitters (also called "tags") are typically purchased as complete units packaged in acrylic or epoxy resin to protect them from weather or the teeth of study animal. Radio-transmitters come as one-stage or two-stage units. One stage transmitters consist of an oscillator and a power source. They are simple, lighter in weight, and have longer battery life for a given tag weight. Two-stage transmitters, have both an oscillator and an amplifier. These tags require more power to operate (>2.4 volts) and are therefore larger, weigh more, have shorter battery life for a given tag weight, and are more expensive. On the other hand, two-stage transmitters have greater range (usually almost double). Animals that are small or have relatively small home ranges are good subjects for one-stage transmitters. Note: It is recommended that the transmitter (with power source) not weigh more than 5% of the subject animal's body weight.

Transmitters are powered by either a lithium or silver battery or by a solar cell. Solar powered transmitters have obvious limitations for studies involving nocturnal, fossorial, or aquatic species. Lithium batteries are the most common source of power. Battery life and signal range are inversely related. Range is generally reported as "line of sight" range and assumes open, level ground with no obstructions. In practice, range is reduced by weather conditions, dense foliage (e.g. tropical forests), uneven terrain, and reflection off water bodies. In general, larger batteries yield increased range and increased battery life, but at the expense of greater weight.

Transmitters can be turned on and shut off by magnetic switches. They are shipped with a magnet taped to the outside of the unit (off configuration) and activated when the magnet is removed in the field. For transmitters with small batteries and short battery life, a magnetic switch may conserve battery life, but they are more expensive and add to the size/weight of the transmitter. Alternatively, transmitters can be shipped with an unfused connection; the biologist solders the connection closed to activate the unit and covers the connection with epoxy in the field.

Each transmitter has a small internal or external antenna attached to the unit. Whip antennas are stainless steel wires coated in Teflon that extend outside the transmitter unit. Whip antennas are light, strong, omni-directional, and produce a relatively uniform signal. In contrast, loop antennas produce their maximum signal when they are tuned to circumference of the subject animals neck (the antenna is generally incorporated into the collar material. Loop antennas produce weaker signals, but are less subject to breakage or chewing.

Radio-telemetry may also be used to provide activity information or certain physiological parameters. Activity sensors use Pulse Interval Modulation (PIM) to represent the animal's activity. For example, real-time PIM sensors use a mercury tip switch to send radio signals of one pulse rate when the mercury in the switch is in one position and a different pulse rate when the mercury bead moves to the opposite position in the switch (e.g. when the animal moves, tipping the switch). Pulse rates that vary tell the researcher that the animal is active. Time delay activity sensors also use a tip switch, but in this case the switch has a counter. The transmitter's pulse rate changes only if the switch is not triggered within a specified period of time. Such sensors are commonly used to detect mortality or tag loss.

Temperature sensors can also be incorporated into transmitters and used to monitor the animal's body temperature. Body temperature transmitters are usually placed subcutaneously or intra-abdominally. As body temperature varies, the pulse rate of the transmitter also varies. Thus it is critical to carefully calibrate the unit ahead of time over a known temperature range. Light transmitters are used to detect the amount of light reaching the transmitter and are typically used to determine activity periods for burrowing species.

TRANSMITTER ATTACHMENT

Transmitters may be attached to wildlife in several ways depending on the animal's body size, shape, lifestyle, and the needs of the researcher. Collar attachment is one of the most popular as it provides a relatively comfortable and durable attachment to the animal (Figure 10.3). Collars should be made of materials such as butyl belts, urethane belts or nylon webbing materials. In certain cases, metal ball-chains or cable ties are also used (small mammals). Properly fitting collars to animals takes some experience. Collars are typically placed so that the transmitter hangs below the neck. They should fit snugly, but not so tight as to be uncomfortable (e.g. hamper breathing). Manufacturers are good sources of information for data on neck circumference in various mammal species. Some manufacturers provide breakaway collars that eventually break and allow the collar to fall off. These collars are useful if the researcher believes that recapturing the animal may be difficult.

Implantable transmitters are placed inside the animals body (e.g. subcutaneously or intra-abdominally). They are typically used for animals with necks thicker than the width of the skull (e.g. small mustelids, mongoose, etc.), for burrowing mammals where a collar may rub against the tunnel walls, or for young mammals that are expected to grow substantially over the course of the study. Implantable transmitters are also used when the study design requires data on body temperature. While implantable transmitters are more comfortable (if implanted properly) to the animal, their range is limited and the subject animal needs to be held in captivity until recovered from the implantation surgery.

Figure 10.3   A radio-transmitter being attached to a moose cow by Scott Becker. (USGS Department of the Interior/photo by Wayne Hubert)

Backpack transmitters are secured to the animal's back by a harness made of plastic-coated wire, nylon webbing, or other materials. The style of backpack harness depends on the study species. Backpacks may snag on vegetation and can cause chafing and are used when other methods are not appropriate. In some cases transmitter assemblies are glued to the animal with cyanoacrylate glue, surgical adhesive, or other substances. Transmitters are typically glued directly to the skin (i.e. the fur is removed over the attachment area) or directly to the fur (e.g., bats, voles). Glued transmitters are positioned so that the transmitter in on the dorsal surface (e.g. the back) with the whip antenna extending toward the tail. These tags generally detach themselves over time as the fur molts or during grooming; this attachment method is generally suitable for shorter duration studies.

There are also a variety of specialized attachment systems including ear-tag transmitters (e.g. for large mammals), and new techniques are in development. Some general rules-of-thumb are to:

· Use the smallest tag possible to accomplish the goals of the study. Generally, less than 5% of the animal's body weight.

· Test the attachment method on captive animals prior to the study.

· Tag at least two animals in large social groups in case one tag fails prematurely.

· Avoid placing tags over markings that the animal uses for intraspecific communication.

· Test each tag prior to attaching it to the animal to ensure it is working properly.

· Follow all guidelines for properly immobilizing animals (or have a veterinarian on hand).

· Allow the animal several days to adjust to wearing the collar prior to collecting data.

· Treat each animal with care during the collaring and tracking process.

RECEIVERS

Radio-tagged animals carry a transmitter that sends out a pulsed signal of a given frequency. The researcher carries a receiver unit that picks up the signal via a hand-held antenna and amplifies the signal so the research can hear it (using earphones). Receivers come in many sizes, weights and prices; the choice will depend on the studies objectives and budget (Figure 10.4). Receivers are powered by batteries (replaceable or rechargeable) and also usually come with a cigarette-lighter adapter for connecting to a vehicle.

Receivers can detect a range of frequencies (e.g. tuned to the transmitter frequency range). Each transmitter frequency is entered into the receiver prior to scanning the area for a signal (Some models have scanners which automatically switch between a number of different frequencies). Like all electronic circuitry, receivers can be damaged by humidity or water, or by excess static electricity (e.g. from vehicle seats and clothing).

Hand-held antennas are attached to the receiver with coaxial cable. The most common antenna is the Yagi or 'H' antenna. These antennas are directional antennas that concentrate their energy toward the front of the antenna (some energy also goes to the sides and behind). The beam width of the antenna is determined by the orientation (vertical or horizontal) and the number of elements in the antenna. For example, a 3-element Yagi antenna will have a narrower bean width if held in a horizontal position (approximately 60 degrees) versus a vertical position. Likewise, this same 3-element antenna will have a narrower beam width than a 2-element antenna held in the same position (roughly 100 degrees). Loop antennas are used for close-range tracking (e.g. 1 km or less). Larger antennas (up to 14 elements) are sometimes mounted on vehicles, boats, or aircraft or used as part of a permanent ground station.

LOCATING ANIMALS

Two methods are typically employed to locate tagged animals. The first, called homing, involves using the receiver and antenna to find the direction in which the signal sounds the loudest. A compass bearing is taken along the line at which the antenna is pointing when the signal is loudest. The researcher follows that line, taking additional readings and adjusting the travel bearing until the animal can be seen.

Figure 10.4 Radio-tracking receiver, headphones, and a 3-element Yagi antenna. (Photo by J. Ryan)

Alternatively, the process of triangulation may be used to infer the position of a tagged animal without direct observation. Triangulation requires finding two or more bearings from different locations and plotting the intersection of those bearings to determine one location point. This process is repeated to collect additional points over time. Each "location" point has an error polygon associated with it (Figure 10.5). The size and shape of this error polygon is determined by a number of factors, including:

· the distance and angle between the bearings when they were taken,

· the distance from the receiver to the tagged animal,

· the bean width of the antenna, and

· the accuracy of the compass bearings.

Ideally, the two bearings should be taken close to the animal and at 90 degrees to each other (Figure 10.5) and a third bearing is usually taken from a third location. The polygon formed by the intersection of the three bearings should be relatively small; the center of the error polygon is considered the animals position at that time. Obviously, if the researcher is alone and has to move from one bearing point to another to take additional readings, then the readings will not be recorded at exactly the same time. If the animal is moving during the time the bearings are taken, then the error polygon will be larger. Consequently, triangulation studies typically rely on stationary base points that are permanently marked and numbered. The personnel at each station should take their bearings at exactly the same time and record them on standard data forms.

Note: The accuracy of a radio-location varies with habitat type and weather conditions. In mountainous areas a common source of error is signal bounce, where a radio signal bounces off a ridge or mountain resulting in huge errors. To reduce sources of error, researchers should ensure that the hand-held antenna is matched to the frequency of the transmitters and held as high above the ground as possible (permanent stations may use poles to elevate the antennae). Alternatively, taking bearings from hill tops or ridge lines may improve the accuracy of the bearings. In addition, it is good practice to take as many bearings as possible, repeat bearings over short time intervals, and get as close as you can to the tagged animal without disturbing it.

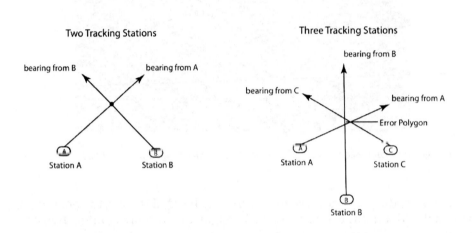

Figure 10.5   Receiver stations are used to establish bearings to the tagged animal. Using two stations (left) yields and estimated position for the animal, but if each bearing is off by a degree or two, then the position changes dramatically. Using three stations simultaneously increases accuracy and reduces the size of the error polygons.

EXERCISES

EXERCISE 1: LOCATING ANIMALS BY HOMING

After you arrive at your designated field site, your instructor will provide each team of students with a radio transmitter, a receiver unit and antenna. Your instructor will familiarize you with the operation of the receiver unit prior to conducting the exercise. Make sure that the receiver is working properly and is receiving the signal from your transmitter.

This exercise simulates how radio-telemetry equipment is used to locate a tagged animal. One student is chosen to be the "*animal*" in the study and the remaining members of the group are the "trackers."

PART I:

The "*animal*" is given a working transmitter and asked to move 100 meters from the "trackers." Each member of the "trackers" should take turns listening for the signal while they scan a full 360 degree circle. In theory the signal should be strongest when the antenna is pointing directly at the "animal". Each "tracker" should practice taking a compass bearing along the line of sight of the antenna's strongest signal. Bearings should be checked with the instructor for accuracy.

PART II:

Once all "trackers" are confident that they can locate the strongest signal and take an accurate compass bearing, the "*animal*" is asked to move out of sight and walk slowly to specific location in the habitat (provided by the instructor). At the designated sight, the "*animal*" places the transmitter on the ground or on a branch of a tree and returns to the "tracker" group. "Trackers" take turns locating the signal, taking a compass bearing, and moving a given number of paces (distance is set by the instructor). The exercise ends when the "trackers" have located the transmitter.

Note: It is very important that you set your compass declination prior to recording any bearings. Magnetic declination is the difference between true north and magnetic north. True north is fixed but magnetic north (the direction the needle of a compass will point) changes over time. Magnetic north is currently some 450 miles from the true north pole. Maps are drawn using true north because it doesn't change. Compasses use magnetic north. In North America, the line of true north and magnetic north are the same at the line of zero declination, which runs through western Lake Superior and across the western panhandle of Florida. West of the line of zero declination, an uncorrected compass will give a reading that is east of true north. The opposite is true if you are working east of the line of zero declination. To correct for magnetic declination, maps typically print the magnetic declination (deviation from true north) to the left of the scale bar (e.g. on a USGS 7.5' quadrangle map). Every time you go into the field you should set your compass declination for the region. If you don't, you will not be travelling in the same direction as on the map. Each compass will differ slightly in how it is corrected for declination, and you should follow the manufacturer's instructions. For example, if the declination at your location is 16° E, you set the declination by rotating the graduated circle on the outside of the compass until it is 16° E from the indicator marker at the top of the compass. To make sure you have set your declination properly, orient your compass so that the north end of the needle is lined up with the 0° mark on the graduated circle. If you are located west of the line of zero declination, then the index pin or marker on your compass should be west of the 0° marker on the graduated circle (and vice-versa if you are east of the line of zero declination). If you have time you can check your declination online at http://www.ngdc.noaa.gov/geomagmodels/Declination.jsp.

Taking a compass bearing is relatively straightforward. If you are using a sighting compass with a diopter sight, you simply hold the compass up to your eye, point the marker on the compass in the direction you want, look through the sighting hole, and read off the bearing in degrees. These compasses are generally accurate to roughly ±0.5°. If you are using a regular baseplate compass, or one with a sighting mirror, then open the cover/mirror and position it at an angle of about 45⁰ above the base and align a landmark in the direction of your bearing (e.g. a tree) with the mirror line. The mirror's vertical line can be seen projected over the compass dial and the bearing read off the compass bezel. A bearing is a measurement of direction and is given in one of two formats, an azimuth bearing or a quadrant bearing. Azimuth bearings use a 360° scale to indicate direction. An azimuth scale compass is numbered clockwise with north as 0°, east 90°, south 180°, and west 270° (e.g. a bearing of 32° would be northeast). Quadrant scale compasses are divided into four 90° quadrants. Azimuth scale compasses are preferred for telemetry studies.

EXERCISE 2: LOCATING ANIMALS VIA TRIANGULATION

In this exercise, you will set up two or more permanent tracking stations and use triangulation to locate the tagged "*animal.*" Your instructor will position the student groups at the tracking stations. Each station is given a letter code. Station locations are determined using a hand-held GPS unit and plotted on a topo map of the area (see Chapter 4 for information on how to use topographic maps). At each station, students mark out the 4 main compass directions on the ground where the antenna will be positioned.

Note: if possible it is helpful to fix the antenna to the top of a tall wooden pole with a 12 inch dowel set perpendicular to the pole (and parallel with the ground). This dowel is set so that it is exactly in line with the antenna set on the pole, and it acts as a general pointing device to indicate the general line of the signal.

With the stations set, students at each station use two-way radios to let the other stations know that they are ready to begin taking readings. The instructor will signal when to be begin and the interval between readings (generally about 3-5 minutes apart). To take a bearing, the antenna (and pole) are slowly turned in a 360 degree circle until the strongest signal is located. A compass bearing is taken along the line of that signal and recorded on data sheets (or in a field journal) along with the time and station code. In this case the tagged "*animal*" is allowed to move, so the bearings will change over time.

Note: It is useful if the student acting as the "*animal*" carries a hand-held GPS unit and records their path as a set of waypoints at each time interval when bearings are being taken by the "trackers." In this way, the triangulation points can be compared to the actual GPS position of the "*animal.*"

At the end of the tracking session, students from each station share data to create a complete set of bearings for all stations at each time point.

TRIANGULATING THE ANIMALS POSITION

The basic principles of triangulation involve locating the receiver stations on a base map and drawing temporary bearing lines from each station. The place where the lines cross is the position (or error polygon) of the animal at the time the bearings were recorded (Figure 10.6).

To begin:

1. Position a sheet of tracing paper over the topo map of the study site and secure it to the map so that it doesn't move.

2. Carefully locate the receiver stations on the base topo map of the study area. Place an x and the station's code letter at each location in pencil.

3. Place a transparent 360⁰ compass circle (Figure 10.6) over the station x so that the center of the circle is at the x.

4. Orient the compass circles so that N on the circles align with N on the map.

5. Beginning with the first set of bearings (e.g. those taken at the same time), draw a faint pencil line that extends from the station x along the bearing (e.g. for a bearing of 43⁰E, the pencil line would begin at station x and extend through the tick mark on the compass circle that corresponds to 43⁰ east (Figure 10.6).

6. Repeat this process for each of the stations using the bearing from that station for that time. Extend each line far enough from the x so that the lines all intersect.

7. At the intersection of all the lines, create a small solid circle at the center of the polygon and write the time next to that point (Figure 10.7).

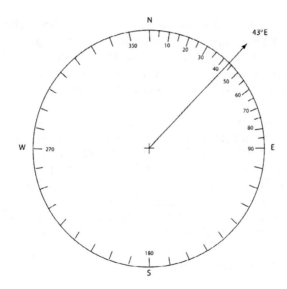

Figure 10.6 An example of a compass circle positioned over a station at X and a bearing of 43 degrees NE.

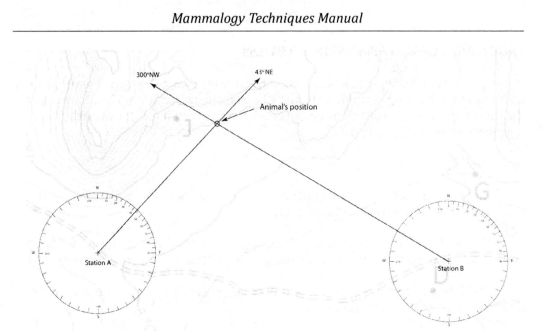

Figure 10.7   Example of two compass bearings taken from two tracking stations. The intersection of the two lines represents the animal's estimated position at that time point.

8. Erase the temporary bearing lines used to create the point (do not erase the point or its corresponding time).

9. Repeat this process until all of the locations have been entered onto the tracing/map paper.

EXERCISE 3: DATA ANALYSIS - THE MINIMUM CONVEX POLYGON

In this exercise we will estimate the home range of the "animals" you tracked using minimum convex polygons (MCP). A minimum convex polygon is one that contains within its perimeter all the line segments connecting any pair of its points (Figure 10.8). Thus, for example, it can not have any indented regions on the perimeter, such a polygon would said to be a concave polygon.

To construct a minimum convex polygon, draw a pencil line around all the perimeter points to create a perimeter with no concave regions. This represents the potential home range of the animals tracked. However, it may not be an accurate representation of the actual home range used by the animal. To see why this is so, look at the location points in Figure 10.8. Notice that many of the points are cluster in a particular subregion of the home range. These points may represent areas of cover, places rich in food resources, or den sites. We will discuss how to deal with this in exercise 4. For now, we will convert the MCP for our data into an area.

The formula for area of a MCP is:

$$A = \frac{x_i\left(y_n - y_2\right) + \sum x_i\left(y_{i-1} - y_{i+1}\right) + x_n\left(y_{n-1} - y_1\right)}{2}$$

where x_i and y_i (i = 1, 2,....n) are the coordinates for the locations.

This can be calculated by hand but it is a difficult task with many location points. There are computer programs that will do the calculations for you and we'll discuss an example in exercise 4. There are other ways to get a reasonable estimate of MCP area by hand. One approach involves plotting the MCP on paper, cutting it out carefully, weighing the cut out MCP, and comparing that weight to another cut out of a known area (e.g. a 100 m² or 1 km²). Try this approach using a photocopy of your MCP and a square of known size (using the same scale as for the MCP).

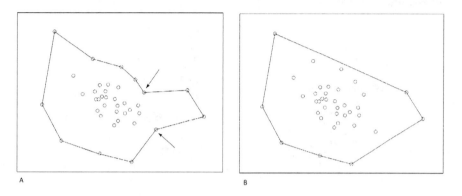

Figure 10.8 Examples of a set of locations bounded by (A) a concave polygon with several indented regions (arrows), and (B) a minimum convex polygon.

A second approach uses a grid whose squares are of known size (Figure 10.9). The MCP is traced onto transparent paper and the MCP is laid on top of and secured to grid. The area of one small square on the grid is determined using the scale on the map, and this area is multiplied by the total number of small squares that fall within the perimeter of the MCP. For example, if one small grid square is 1 km² and there are 340 small squares that fall within the perimeter of the MCP, then the area of the MCP is 340 km². Try this approach using the grid paper provided by your instructor.

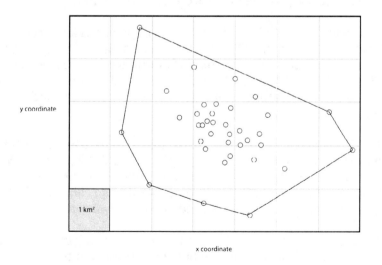

Figure 10.9 A MCP of location points placed on top of a grid, where one grid square is 1km².

EXERCISE 4: DATA ANALYSIS USING LOCOH

There are a variety of methods for estimating home range. As described in exercise 3, the minimum convex polygon approach is by far the most common and the simplest. However, it may not be the best estimate. The minimum convex polygon (MCP) is appropriate only if the locations are relatively uniformly distributed (e.g. the points do not cluster in one or more areas). If location points are clustered in one or more centers of activity, then the MCP method will overestimate the area of the home range (Figure 10.10).

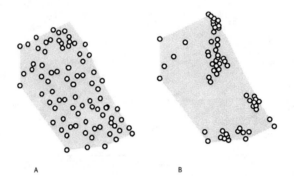

A   B

Figure 10.10   Examples of two identical size home ranges (based on the MCP). In (A) the location points are relatively uniformly distributed across the range, but in (B) the location points are clustered in several centers of activity and there are large areas that are not regularly occupied by the animal.

How can you more accurately estimate the actual home range used by the tagged animal? The answer is to use techniques such as harmonic mean (HM), kernel methods, or local convex hull (LoCoH) models. Recall from exercise 3 that the minimum convex polygon (MCP) is the area of a minimum polygon containing all the location points (e.g. like a rubber band stretched around the outermost points). The other methods are more suitable for most data sets, assuming adequate sample sizes, because animals rarely use their habitats uniformly; some habitat patches contain more resources or provide more protection than others.

Kernel methods and LoCoH methods are more powerful and produce more accurate estimates of home range (The mathematical details underlying these methods are well beyond the scope of this discussion, see work by Getz, et al., 2007; Seaman and Powell, 1996; Worton, 1989). Both MCP and parametric kernel methods have major drawbacks, notably that they are very sensitive to outlying points (Figure 10.11).

Recently, Getz and colleagues have developed a set of new methods based on Localize Convex Hulls, called LoCoH for short (Getz et al., 2007; Getz and Wilmers, 2004). There are three distinct methods in LoCoH: Fixed k LoCoH, Fixed r LoCoH, and Adaptive LoCoH (Getz et al., 2007). Fixed k LoCoH (also called k-NNCH for k-Nearest Neighbors Convex Hulls) will be used in our analysis of home ranges.

The fixed k-LoCoH takes each data point (location) and locates its k-nearest neighbors. It then forms a convex polygon hull using these points (much like the MCP approach, but for a subset of k points). Each hull is successively merged from small to large to form isopleths (contour lines). For example, the 10% isopleth contains 10% of the points and the 100% isopleth contains all points. Therefore, the 100% isopleth is essentially the same as the MCP.

Figure 10. 11 An example of the home range of a wild boar using the Kernel method.

PROCEDURES FOR USING LOCOH

Getz and colleagues provide an online web-based LoCoH calculator (there is also a version available for use with ArcGIS and R project statistical package). We will use the online version for this demonstration.

1. Open your web browser and go to the following URL:

http://locoh.cnr.berkeley.edu/

2. There is a tutorial that describes in more detail the basic features and output of LoCoH. It can be accessed via the link at the top of the web page called "LoCoH Web Tutorial," read it carefully before beginning.

3. Back at the LoCoH Web Application page, you will see an "About" box that gives a very short overview of the procedures. Below that is a pink box "Step 1" that provides 3 options for entering data. By default it is set to use a dataset of wolf sightings from Yellowstone National Park. We'll use this data first. Leave the default "use demo dataset" checked.

4. The next box (yellow) provides a set of options for running the analysis. We are only going to use the fixed K LoCoH so leave the "Fixed K" button clicked.

5. Set the value of k to 400. This means that there will be 400 nearest-neighbor points and essentially this will give us the MCP home range.

6. Leave all the remaining settings at their default values.

7. Click on the *Analyze* button in the Green box to run the LoCoH analysis. This may take a few minutes.

8. When the analysis is complete a new web page will load that displays your results.

9. For our purposes, we are most interested in the top figure of home range coverage (Figure 10.12).

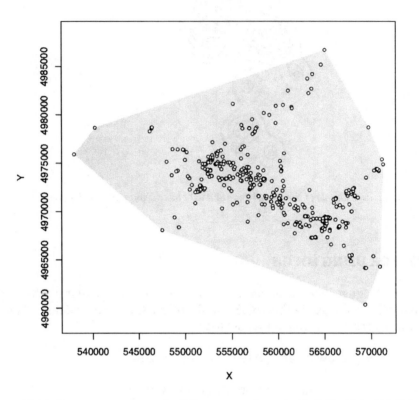

Figure 10.12 Home range coverage for Yellowstone wolves using LoCoH with k=400. This is a single isopleth equal to the minimum convex polygon.

10. Notice that this is essentially a MCP that includes all the data points. It also shows that much of the home range was not used by the wolf. Suppose we want to use LoCoH to generate a more accurate picture of how the wolf uses its home range (e.g. a utilization distribution). Scroll down to the yellow box at the bottom of this page. It allows you to rerun the analysis with different parameters.

11. Set the K value equal to 15 (15 nearest neighbors) and click *Analyze*.

12. The new data page displays the LoCoH results based on a k=15 (Figure

10.13).

13. Notice that with k=15 you have colored polygons where the darker colors represent areas of higher isopleth and therefore higher use by the wolves. It more accurately reflects how the wolves use the area.

Note Suppose you want to rerun the analysis with several new values of k. You can enter more than one at a time (within reason) by separating each value of k by a comma (e.g. enter 5, 10, 15 into the box).

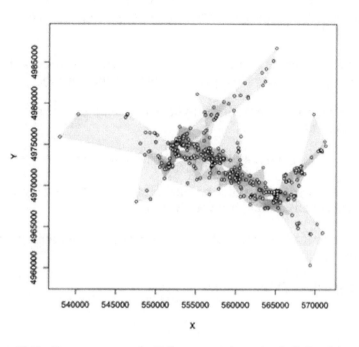

Figure 10.13   Home range map for Yellowstone wolves using LoCoH with k=15.

Note: don't enter more than 3 values of k at a time and do not use datasets with more than about 500 points with the online version - you don't want to crash the server. If you need to use larger data sets you should run Lo-CoH using the R script and R programming language on your own computer - information on how to do that is provided on the web page.

Further down the results page you should see a Summary window and a link called "Download all Data on the Page". Click on this link to save all of the figures and data tables to your computer (rename it something you can recall).

Take a look at the data summary for the wolf data set. You will see a set of areas for each isopleth given as:

```
***Areas of Isopleths:
1.k15
100 199006758.8
90 73582039.3
80 42859709.8
70 27011668.7
60 16124437.5
50 9792377.1
40 5437782.3
30 2931029.1
20 1707891.4
10 492558.1
```

These values are in square meters. Thus, the area within the 100% isopleth (MCP) is 199006758.8 m^2, or 199 km^2. However, the wolves don't use the entire area so a more accurate estimate of their home range might be the 50% isopleth (e.g. approximately 9.7 km^2).

Now you are ready to analyze some of your own data. Your radio tracking data needs to be in x, y coordinates in units of meters (or kilometers).

1. Locate your data file (it needs to be in tab delimited format with no labels at the top of the columns).

2. Return to the main LoCoH web application window.

3. Under *Enter your data* click on the button called *Upload my own tab delimited text file*. This opens a small browser window.

4. Click on *Browse*, navigate to the location of your data file, and click *Open* to load that data set.

5. Set the value of k and run the analysis.

BIBLIOGRAPHY

Getz W. M., Fortmann-Roe, S., Cross, P. C., Lyons, A. J., Ryan, S. J. and C. C. Wilmers. (2007) LoCoH: Nonparameteric kernel methods for constructing home ranges and utilization distributions. *PLoS ONE* 2(2): e207.

Getz, W. M. and C. C. Wilmers. (2004) A local nearest-neighbor convex-hull construction of home ranges and utilization distributions. *Ecography* 27:489-505.

Kenward, R. (1987) *Wildlife Radio Tagging: Equipment, Field Techniques and Data Analysis.* Academic Press, New York.

Nams, V. O. (1989) Effects of radio-telemetry location error on sample size and bias when testing for habitat selection. *Canadian Journal of Zoology*, 67:1631- 1636.

Pollock, K. H., Winterstein, S. R. and M. J. Conroy. (1989) Analysis of survival distributions for radio-tagged animals. *Biometrics*, 45:99-109.

Seaman, D. E., Griffith, B. and R. A. Powell. (1998) KERNELHR: a program for estimating animal home ranges. *Wildlife Society Bulletin,* 26: 95-100.

Seaman, D. E. and R. A. Powell. (1996) An evaluation of the accuracy of kernel density estimators for home range analysis. *Ecology*, 77:2075-2085.

Worton, B. J. (1989) Kernel methods for estimating the utilization distribution in home-range studies. *Ecology*, 70:164–168.

11

GPS Tracking Using *Google Earth* and *MoveBank*

TIME REQUIRED

One typical 3-4 hour lab period.

LEVEL OF DIFFICULTY

Easy

LEARNING OBJECTIVES

Understand how the Global Positioning System is used to track animals.

Learn the format for GPS coordinates.

Use *Google Earth* for analyzing GPS data.

Use *GPSVisualizer* online software to convert file types and plot GPS data.

Access and analyze telemetry data sets from *MoveBank*

EQUIPMENT REQUIRED

Computers with access to the Internet

Data sets of GPS coordinates

BACKGROUND

In radiotelemetry a tagged animal wears a radio transmitter that sends out a radio frequency signal that can be picked up by a researcher carrying a radio receiver tuned to that frequency (see Chapter 6). This gives the approximate direction to the animal, but only with triangulation from several locations can the animal's approximate position be determined. Radiotelemetry is a relatively inexpensive way to track animals over relatively short distances. Researchers wishing to track large mammals over long distances (e.g. seasonal migrations) or from more remote sites now use GPS and/or satellite telemetry systems. The "Global Positioning System" (GPS) uses a series of 24 to 32 geostationary satellites in medium-Earth orbit that send signals to GPS devices on the ground. Using information from a minimum of four satellites, the GPS unit calculates the position on the ground (latitude, longitude, and altitude) plus the current time. GPS units can be attached to animals and the location data stored on the GPS collar or transmitted to the researcher via satellite or cellular phone signals. This means that a researcher half a world away can track the movements of a tagged animal in near real-time (Hebblewhite and Haydon, 2010; Handcock et al., 2009; Tomkiewicz et al., 2010).

An animal carrying a GPS-enabled tag can record and store location data at a pre-determined interval (e.g. every hour or twice a day). These data are stored in an onboard microprocessor and the data are recovered when the animal is recaptured and the collar removed. Alternatively, GPS collars can be retrieved via remote-release systems that detach the collar automatically. Remote-release collars send out a radio signal upon release and are located in the field using radiotelemetry.

More expensive GPS tags are coupled with satellite transmitters that relayed the units position information to a satellite and from there to a central data center on the ground (Figure 11.1). For example, an animal with such a collar would receive a signal from GPS satellites, calculate its three dimensional position on the Earth, transmit the location to an "Advanced Research and Global Observation Satellite" (Argos), which would then relay that information to ground-based data processing center for relay onto the researcher. The researcher plots the animal's location on a map in near real-time using geographic information system software (GIS).

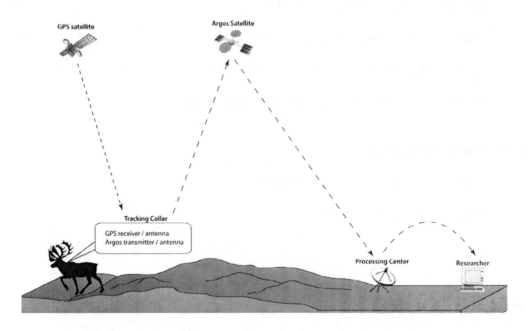

GPS satellite

Argos Satellite

Tracking Collar

GPS receiver / antenna
Argos transmitter / antenna

Processing Center

Researcher

Figure 11.1   A diagram of the GPS-Argos system for remotely tracking large mammals.

In the nearly 50 years since the Craigheads (Craighead, 1982; Craighead et al., 1995) first radiotracked grizzly bears in Yellowstone National Park, there has been a radical change in how ecologists study the distribution and movement patterns of mammals (both terrestrial and marine). As Hepplewhite and Haydon (2010) point out, "today, ecologists sitting at their desk can check the movements of even the most difficult to study species such as GPS radiocollared wildebeest (*Connochaetes taurinus*) or Argos-tagged bluefin tuna (*Thunnus thynnus*) on *Google Earth*, as they check their morning email with minute-by-minute data streams."

EXERCISES

EXERCISE 1: TRACKING GRIZZLY BEARS WITH

GOOGLE EARTH AND *GPSVISUALIZER*

This exercise uses GPS data for grizzly bears from North America. The data is in spreadsheet format. The data file was generously provided by Dr. Rick Mace at Montana Fish, Wildlife and Parks and can be downloaded at:

http://www.wildmammal.com/downloads

To begin, Open the file "Griz-June/Oct 2007.xls" in *Excel* or *OpenOffice*. Look at how the data are formated (Figure 11.2). The data is separated out by month with the June data in one worksheet and the October data in another (see the tabs at the bottom of the spreadsheet). The main data we are interested in is the latitude and longitude, which will allow us to fix the animals position in space and time.

MONTH	DAY	YEAR	gmt	GMT_HR	DEAD_ALIV	lat	long
6	13	2007			alive	48.87068	-113.909281
6	14	2007	0.00154	0	alive	48.86902	-113.906891
6	15	2007	0.00076	0	alive	48.86997	-113.900202
6	15	2007	0.50076	0.5	alive	48.87483	-113.897089
6	15	2007	0.75044	0.7	alive	48.86923	-113.906337
6	16	2007	0.00096	0	alive	48.87525	-113.905477
6	16	2007	0.50076	0.5	alive	48.86898	-113.895756
6	16	2007	0.75043	0.7	alive	48.87457	-113.897421
6	17	2007	0.0011	0	alive	48.87528	-113.90377
6	17	2007	0.75064	0.7	alive	48.87909	-113.900421
6	18	2007	0.00075	0	alive	48.87535	-113.902149
6	18	2007	0.25179	0.2	alive	48.87574	-113.902975
6	18	2007	0.50076	0.5	alive	48.87767	-113.900963

Figure 11.2 Format of the data in the grizzly data set.

In order to display these coordinates visually, you will need to map it. To do this you will take advantage of two web-based software programs: *Google Earth* and *GPSVisualizer*. *Google Earth* is a powerful tool for displaying data on satellite images of the entire world. However, *Google Earth* can not read raw spreadsheet data (such as .xls files from *Excel*). The first step is to convert the raw spreadsheet data into a format that *Google Earth* can display.

1. Open your web browser and navigate to the *GPSVisualizer* website at:

http://www.gpsvisualizer.com/. This will open a page like the one show in Figure 11.3.

2. You need to upload some data for *GPSVisualizer* to work with. In the *Get Started Now* box click the *Browse* button and locate the file "Griz-June/Oct2007.xls" on your computer. Select your data file and click *OK* to enter it into the box next to "Upload GPS file."

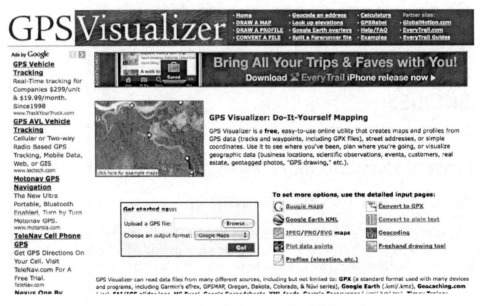

Figure 11.3 Screen shot of the main page of GPSVisualizer.

3. Click the arrows on the right of the *Choose and output format* box to get a list of choices – click on *Google Earth* and then click the *Go* button. You will see a page like the one in Figure 11.4. Your converted file now ends in .kmz.

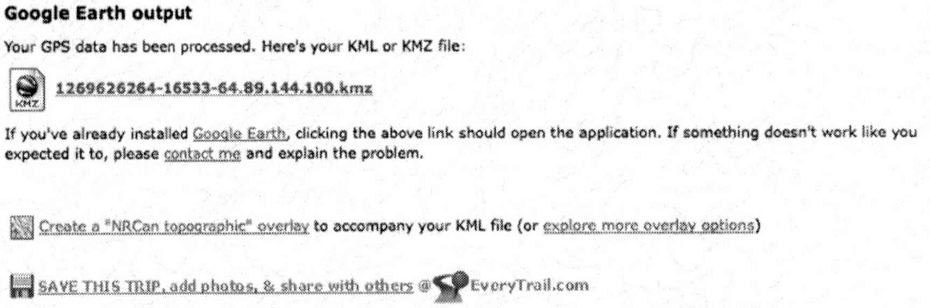

Google Earth output

Your GPS data has been processed. Here's your KML or KMZ file:

1269626264-16533-64.89.144.100.kmz

If you've already installed Google Earth, clicking the above link should open the application. If something doesn't work like you expected it to, please contact me and explain the problem.

Create a "NRCan topographic" overlay to accompany your KML file (or explore more overlay options)

SAVE THIS TRIP, add photos, & share with others @ EveryTrail.com

Figure 11.4 The output of GPSVisualizer. The spreadsheet data has been converted into a format (.kmz) that Google Earth can read.

4. To open this file click on the file name (ending in .kmz) to open a box like the one in Figure 11.5. It's a good idea to save this new file first. Click the *Save File* button, then click *OK* (save it to a convenient location). Go back to *GPSVisualizer* and open your .kmz file in *Google Earth* by clicking the *Choose* button and locating your copy of *Google Earth* (typically it is in your Applications folder (Macs) or in the Programs

Folder in Windows), and click on your *Google Earth* icon to launch this .kmz file in *Google Earth*.

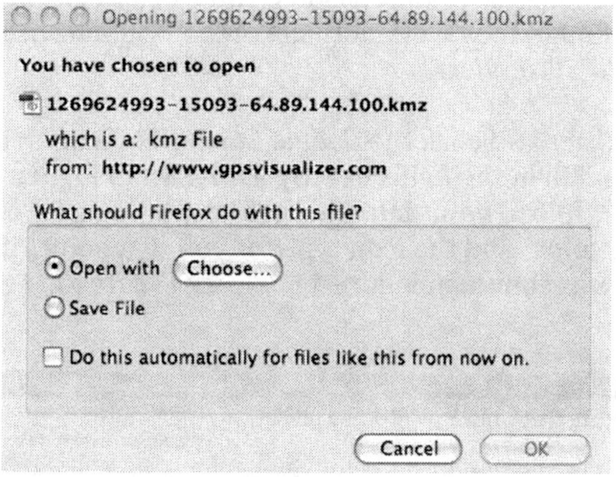

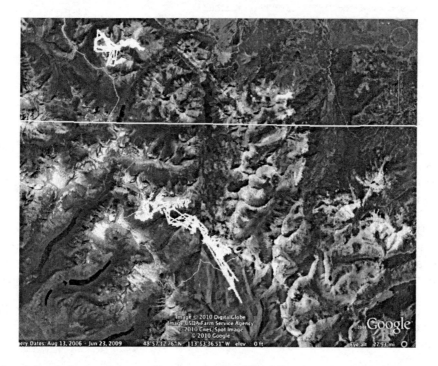

Figure 11.5   GPSVisualizer will open your .kmx file in Google Earth.

5. Google Earth will open with the data displayed. On the far left of *Google Earth* you should see a panel that provides additional features and will also show your data file under "Temporary Places."

6. Explore your data for this female grizzly by zooming in and adjusting your observation point using the navigation tools in the upper right corner of the *Google Earth* display (Figure 11.6). Notice that this female spent June in Glacier National Park in the US, and October in Waterton Lakes National Park, Canada.

Figure 11.6   Google Earth display of the tracking data for June (bottom) and October (top) overlayed on a satellite map of the region. The horizontal line is the US-Canada boarder.

7. One of the things we can see from this map is that the area is mountainous. You might want to know at what elevations the female grizzly spent most of her time, and did the elevation change with the seasons. You can add elevation information to the GPS points even though they were not part of the original spreadsheet data. To do this we again use *GPSVisualizer.*

8. Return to *GPSVisualizer* and click the *Google Earth* KLM icon to the right of the "Get started now" box. Fill in the fields exactly as shown in Figure 11.7 (except add your data path in the "Upload your data file here" box). You may change a few settings such as "default color" and the icon type for you waypoints if you like, but leave the rest of the settings as shown in Figure 11.7. Click *Create KLM* file button.

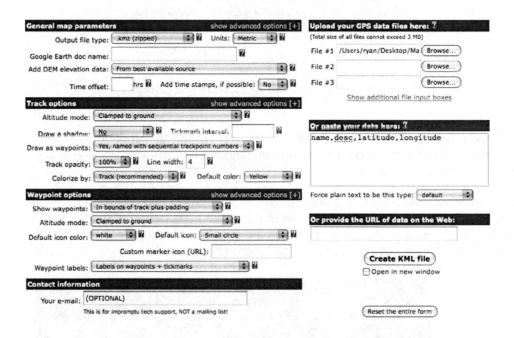

Figure 11.7   The Google Earth KLM window. Set you parameters to match those shown here (except you will upload you own file).

9. Click on the KLM file that appears in the gray box (as you did before), select the *Choose* button, and locate your *Google Earth* application, then click on it and click *Open* button and then the *OK* button. *Google Earth* reloads the data, but now with the elevation data attached.

10. Drag your mouse over any waypoint on the map and/or click on one waypoint to call up a balloon with the elevation data for that point (Figure 11.8). You now have elevations assigned to each waypoint. However, you might like to export that data to your spreadsheet so you can calculate mean elevations and other parameters.

11. Return to *GPSVisualizer* main page and click on the button called *Profiles* (elevation etc.). It will open another screen called draw a profile. Fill in the data boxes as in Figure 11.9.

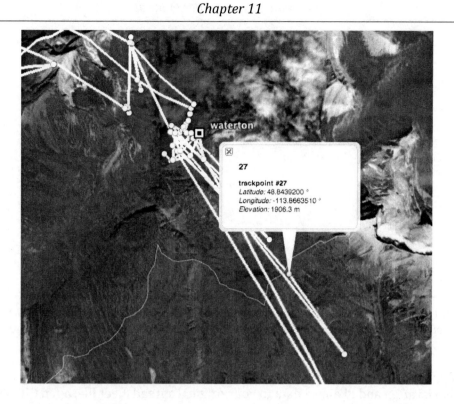

Figure 11.8   Grizzly tracking data with one waypoint highlighted to show its coordinates and elevation.

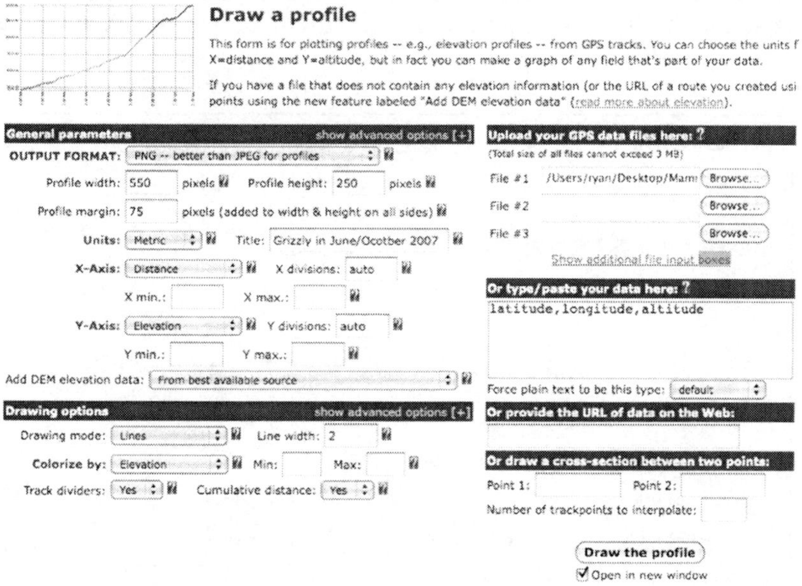

Figure 11.9   The draw a profile data entry window in GPSVisualizer. Fill it in exactly as shown here.

12. Click the *Draw the Profile* button at the bottom right. You will see something like the plot in Figure 11.10. This plot shows the altitude of the grizzly for June and October. Notice that there is a sharp increase in altitude after the animal travelled approximately 80 kilometers. This is an artifact produced because the July to September data were deleted. Download your plot data by clicking the green *download* link in the top line of text.

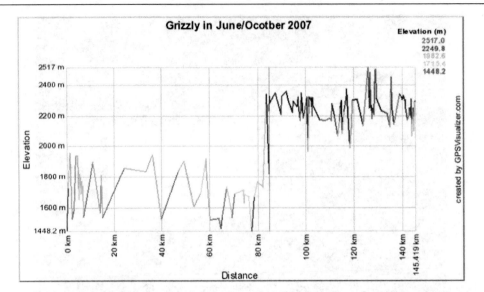

Figure 11.10 A plot of the elevations vs distance traveled by the female grizzly in June and in October. Notice that there is a distinct break at the vertical line to show the June versus October data.

13. Suppose you want to store that elevation data with your original lat/long. data. To add elevation and distance data to your original spreadsheet file, open GPS Visualizer and click on *Convert to Plain text* in the list of options to the right. You'll get a gray data entry box. Fill it out as in Figure 11.11 and add your data file in the *Upload your files here* using the *Browse* button. This will add distance, speed, and elevation.

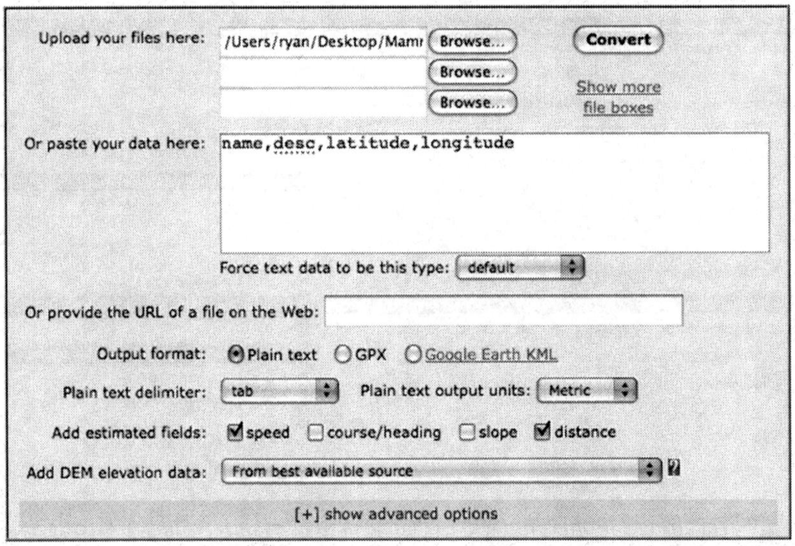

Figure 11.11 The data entry box for converting a file to plain text in GPSVisualizer.

14. Now click the *Convert* button to get a txt file like the one shown in Figure 11.12. For example, the first entry in this file occurred on June 13th, and the new data on altitude (1451.4 meters) and distance (0.00 meters) are added to the spreadsheet. Notice that the distance is 0 because this is the starting point.

The contents of your file are also displayed in this box, if you'd rather cut and paste:

type	year	month	day	latitude	longitude	altitude (m)	distance
T	2007	6	13	48.8706800	-113.9092810	1451.4	0.000 Oct 2007
T	2007	6	14	48.8690200	-113.9068910	1507.9	0.255
T	2007	6	15	48.8699700	-113.9002020	1730.5	0.757
T	2007	6	15	48.8748300	-113.8970890	1959.5	1.343
T	2007	6	15	48.8692300	-113.9063370	1526.7	2.264
T	2007	6	16	48.8752500	-113.9054770	1588.5	2.937
T	2007	6	16	48.8689800	-113.8957560	1939.0	3.934
T	2007	6	16	48.8745700	-113.8974210	1938.9	4.568
T	2007	6	17	48.8752800	-113.9037700	1655.3	5.040
T	2007	6	17	48.8790900	-113.9004210	1821.5	5.530
T	2007	6	18	48.8753500	-113.9021490	1721.7	5.964
T	2007	6	18	48.8757400	-113.9029750	1688.6	6.039

Map this data: Google Maps, Google Earth, JPEG map, SVG map, or elevation profile -- or go to the map form to set options

Figure 11.12   The new text file with the altitude added to the original latitude and longitude data in GPSVisualizer.

15. Just above the data file is a green link called *click to download ....txt*. Click on this link to download and save your text file to your desktop.

16. You can also click the *Map this data* link below the sample table to instantly map the data.

17. After you have downloaded the text file with the elevations to your computer, you can import this .txt data file into *Excel* or another spreadsheet program (Figure 11.13).

	A	B	C	D	E	F	G	H	I	J
	type	year	month	day	latitude	longitude	altitude (m)	distance (km)	name	dead_al
	T	##	6	13	48.87068	-113.90928	1451.4	0	Oct-07	alive
	T	##	6	14	48.86902	-113.90689	1507.9	0.255		alive
	T	##	6	15	48.86997	-113.9002	1730.5	0.757		alive
	T	##	6	15	48.87483	-113.89709	1959.5	1.343		alive
	T	##	6	15	48.86923	-113.90634	1526.7	2.264		alive
	T	##	6	16	48.87525	-113.90548	1588.5	2.937		alive
	T	##	6	16	48.86898	-113.89576	1939	3.934		alive
	T	##	6	16	48.87457	-113.89742	1938.9	4.568		alive
	T	##	6	17	48.87528	-113.90377	1655.3	5.04		alive
	T	##	6	17	48.87909	-113.90042	1821.5	5.53		alive
	T	##	6	18	48.87535	-113.90215	1721.7	5.964		alive
	T	##	6	18	48.87574	-113.90298	1688.6	6.039		alive
	T	##	6	18	48.87767	-113.90096	1775.6	6.299		alive
	T	##	6	18	48.87636	-113.90265	1695.8	6.491		alive
	T	##	6	19	48.87501	-113.90159	1743.7	6.66		alive
	T	##	6	19	48.877	-113.90682	1540.1	7.103		alive
	T	##	6	20	48.8514	-113.8731	1901	10.875		alive

Figure 11.13   The new spreadsheet with altitude and distance for each point.

One final note that may be worth while is that you can create a tour of your way points (similar to a flyover). You can do this by selecting the appropriate data in My Places (on the left hand side of the *Google* interface. To do this you need to find the appropriate line (path) in My Places, select the appropriate line in the Places panel and click the *Play Tour* button (located at the bottom of the Places panel (it is an icon with three small polygons next to the slider).

The tour begins playing in the 3D viewer and the tour controls appear in the bottom left corner of the 3D viewer. To pause or resume the tour, click the *Pause/ Play* button. To fast forward or go back on the tour, click the arrow buttons (press these repeatedly to accelerate back or forward). To replay the tour, click the *Repeat* button. Use the tour slider to move to any part of the tour.

These controls disappear if the tour is inactive for a period of time, but you can make them reappear by moving the cursor over the bottom left corner of the 3D window.

QUESTIONS:

1) What is the average rate of movement per day for each month for this female grizzly?

2) What is the average elevation change per day in each month? What might explain the differences in elevation?

3) What is the maximum elevation change and maximum distance moved in one day (for each month)?

4) Do grizzly bears prefer dense forest or more open areas? This requires you to zoom in on data points in *Google Earth* and inspect the habitat.

5) Do grizzly bears prefer north or south facing slopes?

6) What is the distance to nearest road, town, water? You will have to click on the check boxes in the layers panel in *Google Earth* to see roads and other features added to your map. Compare June (just after arousal from hibernation) with October (just prior to entering hibernation).

EXERCISE 2: EXPLORING *MOVEBANK* DATA

In this exercise we will use an online database of telemetry information to access and display GPS data for African buffalo in Kruger National Park, South Africa. The data set can be found online at *MoveBank:*

http://www.movebank.org

MoveBank is a community of researchers who have uploaded their personal data sets to a database to allow broader access to tracking data. *MoveBank* acquires new data by linking to satellites (or cellphone networks, etc), and uploads the data to a centralized database. Users can then display or analyze these data using *MoveBank*'s website.

If you become a regular user of *MoveBank*, you should register with the site. Go ahead and do this now. Login again using the username and password you have just created.

Note: You may use the data sets in *MoveBank*, but you may not publish the results without permission from the original owner (collector) of the data set.

---

1. Open the *MoveBank* page on your browser at http://www.movebank.org. You should see the main data search page (Figure 11.14). The main page displays a data searching panel on the left and a *Google* map on the right. The locations of the

datasets in the database are displayed on the map with gray or green dots. The map can be set to display as a basic "road map", a satellite image map, or a terrain map by clicking on the buttons in the upper right side of the map. Try switching between map settings, but return to *Terrain* before continuing with this exercise.

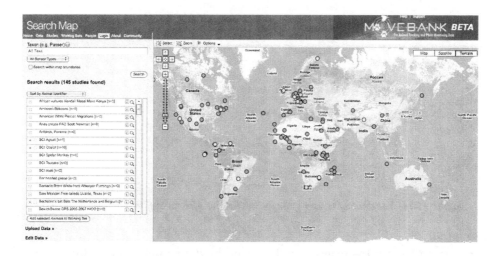

Figure 11.14   MoveBank's main home page displays a Google map on the right and a data searching panel on the left.

2. Using the data searching panel on the left, scroll down the list of data sets until you locate the listing for "Kruger African Buffalo, GPS tracking, South Africa".

3. Click in the small box immediately to the left of the name of the data set (Figure 11.15) to select this data to work with.

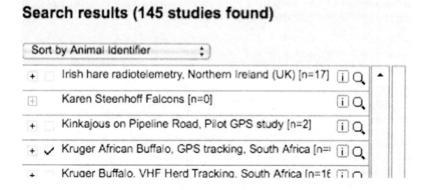

Figure 11.15   Close up of the data search panel showing the Kruger African Buffalo data set is selected with a check in the left hand box.

4. Click on the small icon on the right side of the data set's name -it looks like a small box with an *i* inside. This opens a popup screen with three choices.

5. Click on the choice *Open in Studies Page.* This will open a new page (Figure 11.16) that provides a summary of the contents of the data set you will work with.

STUDY DETAILS

| Summary | Map | Sharing | Download |

Study Name	Kruger African Buffalo, GPS tracking, South Africa
Contact Person	Paul Cross
Principal Investigator	Paul Cross
PI Contact Details	2327 University Way, Suite 2 Bozeman, MT 59715
PI Email	pcross@usgs.gov
Citation	*not set*
Acknowledgements	Collection of Kruger Park Buffalo data funded by NSF Grant DEB-0090323 to Wayne M. Getz
Grants used	NSF Grant DEB-0090323 to Wayne M. Getz
License Terms	Do not use for any publication without consulting Paul Cross
Study Objectives	*not set*
Study Reference Location	
Longitude	31.752
Latitude	-24.389
Study Statistics	
Number of Animals	6
Number of Tags	6
Number of Deployments	6
Number of Locations	30094
Time of first location	2005-02-17 05:05:00.000
Time of last location	2006-12-31 14:34:00.000
Sensor Types	GPS
Taxa	Syncerus caffer

Figure 11.16   The data summary page for the Kruger African buffalo data set in MoveBank.

6. This page displays the original collector of the data and his/her address, the license terms, the number of tagged animals (6), the number of locations (30,094), and other relevant information. Notice that this is a very large data set.

7. Use your browser's *Back* button to go back to the main data search page.

8. Click on the small *magnifying glass* icon next to the Kruger data listing in the search panel (Figure 11.15). This displays a terrain map zoomed in to the region in South Africa where the data can be seen as a series of pink patches (actually pink points).

9. Click on the center of the map and hold down the mouse button to drag the map until your data are in the middle (Figure 11.17).

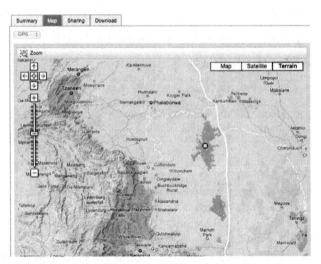

Figure 11.17   Terrain map of the African buffalo. Notice that they form two separate herds.

10. This map shows two distinct herds of African buffalo, one to the north and one to the south. The terrain map does not show the boundary of Kruger National Park, so it is difficult to know if the herds are located within the park or sometimes cross park boundaries.

11. Click on the *Map* button in the upper right to change to the basic "road map" view. You can now see the park boundary. Click on *Satellite* button to get a photo image. Note how the buffalo move relative to the large rivers in the area.

12. Switch back to the *Terrain* view.

13. Click on the *Options* button at the top left of the map. You will see a drop down menu. Click the box for *Draw lines for selected animals* and uncheck the box for *Draw lines for highlighted animals*. Then click *Close* to close the drop down menu.

14. Use the map zoom features on the left hand side of the map to zoom in on the two buffalo herds. You can do this by clicking the + button several times (and re-centering you map) or you can drag the zoom slide bar up until it is in the fourth position down from the top +. You will now see many tiny pink dots representing the GPS locations for one buffalo at a specific time point. You should also see path lines between points with arrows pointing in the direction of movement.

15. Click on one path line (zoom in further if needed). This displays the points for one of the 6 African buffalo in blue (Figure 11.18). Now you can see how this individual moved relative to the other tagged buffalo in that herd.

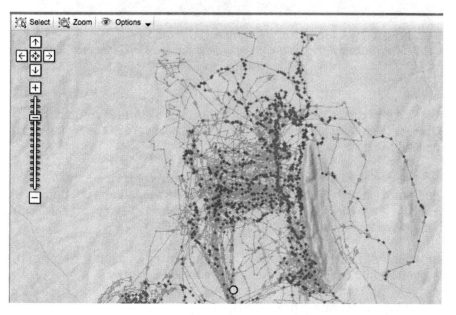

Figure 11.18 Close up map of the African buffalo herd with one individual selected.

16. Click back and forth between the *Satellite* and *Terrain* views to see how the herd moves relative to geographic features (e.g. steep ridges, open savanna, water sources, etc).

17. Click on the small *i* icon on the right hand side of the data set listing (Figure 11.15) to display a popup of choices for what to do with the data.

18. Click on the *Open in Google Earth*. Agree to the terms of use.

19. Open the resulting kmz file in *Google Earth* (as described in exercise 1), or save the kmz file to your computer for later use.

20. Google Earth displays the African buffalo data set (Figure 11.19).

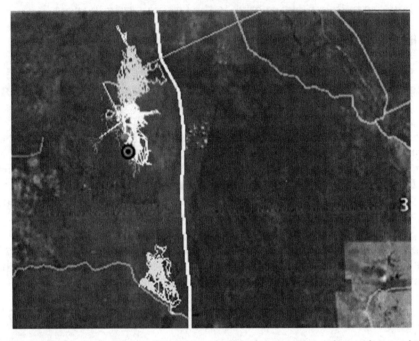

Figure 11.19 Google Earth display of the African buffalo data set. Notice that each animal is given a unique color.

21. Notice that you can also download the data set as a csv file (comma separated values) for use in spreadsheets or other software. Be aware of the terms of use if you download the data.

22. Select a second data set to explore. Use the features described above to explore the animal's movements. For example you might use the Indiana bat data to explore how these bats leave their hibernacula, or explore how urban coyotes use suburban and urban environments at night.

23. Write a detailed description of how this animal uses its habitat. Some of the questions you should consider in you write up include:

    – What type of terrain is preferred?

    – What type of cover is used?

    – Do movements change with time of day or night?

    – When is the animal most active?

    – How far do the animals typically travel in a day?

– What altitude do they prefer?

– Do they avoid human-modified landscapes? If so, how?

– How do their movement patterns relate to water sources?

For some of these questions you may need to export the data to *Google Earth* or as a csv file that you open in a spreadsheet.

BIBLIOGRAPHY

Craighead, F. C. (1982) *Track of the Grizzly*. Random House, NY.

Craighead, J. J., Sumner, J. S. and J. A. Mitchell. (1995) *The Grizzly Bears of Yellowstone: Their Ecology in the Yellowstone Ecosystem.* Island Press, NY.

Handcock, R. N., Swain, D. L., Bishop-Hurley, G. J., Patison, K. P., Wark, T., Valencia, P., Corke, P. and C. J. O'Neill. (2009) Monitoring animal behaviour and environmental interactions using wireless sensor networks, GPS collars and satellite remote sensing. *Sensors* 9, 3583–3603.

Hebblewhite, M. and D. T. Haydon. (2010) Distinguishing technology from biology: a critical review of the use of GPS telemetry data in ecology. *Philosophical Transactions of the Royal Society B*, 365:2303-2312.

Tomkiewicz, S. M., Fuller, M. R., Kie, J. G. and K. K. Bates. (2010) Global positioning system and associated technologies in animal behaviour and ecological research. Philosophical Transactions of the Royal Society B, 365, 2163–2176.

12

Recording and Analyzing Mammal Sounds

TIME REQUIRED

Approximately one 3 to 4 hour lab period for each exercise.

LEVEL OF DIFFICULTY

Moderate

LEARNING OBJECTIVES

Understand the role vocalizations play in animal communication

Understand the importance of alarm calls in animal societies

Learn how to record animal vocalizations in the field

Understand how sounds are displayed visually in a spectrogram

Learn how to measure and manipulate audio information from a spectrogram

Apply knowledge of sound and communication to a field experiment

Conduct a playback experiment

EQUIPMENT REQUIRED

Binoculars

Digital recording device (using SD memory cards)

Field microphone (shotgun or parabolic)

Battery powered field speakers

Audio analysis software (Audacity or Raven Lite)

Data sheets and clip boards

BACKGROUND

Mammals have an acute sense of hearing, and auditory communication is very important in most mammal species (Vaughan et al., 2010). Mammals use acoustic signals to maintain contact with other members of the herd or pack, as part of courtship behaviors, to warn of danger, and to recognize individuals within a colony or social group (e.g. mother-infant communication). Primates and cetaceans in particular exhibit complex repertoires of acoustic signals (Goodall, 1986; Weilgart and Whitehead, 1997). With the advent of more sophisticated sound recording equipment, researchers demonstrated that mammalian auditory communication extends well above (ultrasonic) and well below (infrasonic) the human hearing range (see Vaughan et al., 2010).

Ultrasonic signals, or echolocation signals, have been known in bats and dolphins for decades. However, it is only in the last decade that researchers have discovered that some mammals communicate over long distances in the infrasonic range (e.g. below 20 Hertz). African elephants produce intense, extremely low-frequency vocalizations (14 to 24 Hertz) that may carry over six miles (Payne et al., 1986). These varied infrasounds communicate locations of individuals or groups across vast distances; they also relay information on reproductive condition, social identity, and alarm (Poole et al., 1988; McComb et al., 2003).

In another recent and fascinating example, researchers discovered that some fossorial (living primarily underground) mammals communicate using a combination of low-frequency vocalizations and seismic signals (vibrations transmitted through the soil). For example, banner-tailed kangaroo rats (*Dipodomys spectabilis*) communicate by footdrumming (Randall, 1997; Randall and Matocq, 1997). Apparently, these kangaroo rats can recognise strangers and neighbors by their unique footdrumming signatures (Randall, 1989).

Mammalogist know relatively little about acoustical communication in most wild mammals. Therefore, we can expect many exciting new discoveries in the future and mammalian bioacoustics remains a promising area for research.

EQUIPMENT FOR RECORDING SOUNDS

The basic equipment to study animal sounds should include:

- microphones, hydrophones, and/or bat detectors (specialized equipment for the recording of ultrasounds)

- a digital (or analog) recorder capable of covering the animal's entire frequency range

- computer software for sound analysis

- digital audio playback system with battery-powered speakers

MICROPHONES

The microphone converts sound pressure into an electrical signal, which can be amplified, recorded, and analyzed. Choosing the right microphone is critical and there are a number of designs on the market. There are five main factors to consider:

· type of transducer,

· efficiency (or sensitivity),

· self-noise (its intrinsic noise),

· frequency response (range of frequencies received), and

· directionality (polar pattern).

Microphones are typically classified by their transducer design, (e.g. condenser or dynamic) and by their directional characteristics. Dynamic microphones are very reliable, more resistant to moisture, and don't require powering. However, they are not as sensitive as condenser microphones.

Condenser microphones have extended frequency response, but require powering (usually via 48 volt Phantom Powering from the recording unit). An electret condenser microphone is a relatively new type of miniature microphone found in most consumer electronics today. Some of these rival traditional condenser microphones in signal quality, yet are small and very inexpensive.

Another important consideration for field recording is directionality. Among directional microphones, the most useful in bioacoustic recordings, are directional microphones (e.g. shotgun microphone) and parabolic microphones. Directional and parabolic microphones collect sound coming directly in front of the microphone and thereby reduce ambient noise (e.g. wind and human activities). Wind noise is reduced by adding special wind shields to shotgun microphones. Parabolic microphones use a dish-shaped parabola to focus incoming sounds on a focal point where a microphone is positioned. Parabolic microphones are somewhat limited by the sounds wavelength. A low frequency sound of approximately 100 Hz would require a parabolic dish three meters in diameter (not a practical field option).

Self noise is also an important consideration for field recording. The best (quietest) have a self noise of 5 to 10 decibels (dB) as compared to the average microphone's 20 to 25 dB. Ideally, one would chose a directional microphone with the highest sensitivity, lowest self-noise, and widest frequency range.

Note: Hydrophones are special transducers that acquire underwater sounds (pressure waves) and convert them into an electric signal. Hydrophones use piezoelectric elements that produces an electrical current when compressed by a sound wave. They are usually omnidirectional and have a frequency response from 2-10 Hz to more than 100 kHz.

DIGITAL AUDIO RECORDERS

Audio recorders can be analog or digital. They allow recording an electrical signal generated by a microphone or hydrophone. Traditional cassette and open reel tape recorders (analog) degrade the signals by adding hiss or other types of distortion. Digital recorders eliminate these problems and record highly accurate, low-noise signals. Additionally, digital audio files can be easily stored, searched, processed using computers. Thus, digital audio recorders are preferred by wildlife researchers (analog systems will not be described here).

DAT (Digital Audio Tape) recorders can still be found on the market, but are being rapidly replaced by solid state recording units. DAT has a frequency response of 10Hz-22kHz and can not be used for ultrasonic signals. MiniDiscs (MD) recorders are still available but they use compression algorithms that degrade the sound (they often delete audio information outside the range of human hearing). The current state-of-the-art in digital recorders are units that record on internal hard-disks or on removable memory cards (e.g. SD memory cards). These units can record in compressed or uncompressed formats depending on the duration or quality needs of the researcher. Another advantage is that some models can record up to 192kHz, although 96kHz is more common. Most are pocket-sized and have microphone preamplifiers. For a comparison of the recorders, visit the web pages of the *Wildlife Sound Recording Society* at

http://www.wildlife-sound.org/equipment/newcomersguide/index.html.

Infrasonic and ultrasonic recorders are specialized devices capable of recording signals whose frequencies are lower or higher than those audible to humans. Ultrasound (echolocation) recordings can be played back at slower speed to make them audible to humans. Devices capable of detecting and recording ultrasound exist to study echolocation in bats (e.g. bat detectors). Three main types of bat detectors are available:

· heterodyne frequency shifting,

· frequency division

· time expansion

Heterodyne detectors shift a small frequency range (several kHz) down to the human audible range. Unfortunately anything outside that frequency range is lost. Frequency division (FD) bat detectors convert the bat's call into a square wave and then divide that by 10 to generate an new square wave. Frequency division detectors cover a very wide frequency range, but square waves sound harsh and contain unwanted harmonics. Heterodyne and FD systems allow recording of ultrasonic signals transformed to audible signals, not the original ultrasonic signal. The time expansion detector (TED) is the most accurate. All the information of the original signal is retained. The bat's ultrasonic signal is digitally sampled at high speed, stored to memory, and replayed at one tenth the rate to make it audible (and recordable). For additional details see (Ahlen and Baagøe, 1999; and the website:

http://www.bats.org.uk/pages/bat_detectors.html

Remember, if you plan to record mammal sounds in the field, you will need a digital recorder capable of:

· Long battery life and ability to use normal or rechargeable AA or AAA batteries.

· AC Power supply capable.

· Case size/button size should be portable but easy to use in the field.

· 1GB memory capacity (or removable memory cards).

· Manual control of record level.

· Bright easy to see record level meters.

· XLR microphone connectors for attaching external mics.

· 1/4 inch phono jack for attaching headphones.

· +48V phantom power (for external mics).

· A sample rate of at least 48kHz and a 24 bit depth.

· An internal speaker or the capacity to connect an external speaker to the recorder for playbacks.

SOFTWARE FOR ANALYZING SOUNDS

Sound analysis software allows users to:

· graphically display acoustic signals

· understand and measure their structure

· correlate it to observed species, behaviours and situations.

Note: This exercise assumes that readers are familiar with basic terms such as frequency, amplitude, wavelength, hertz, and decibels. If not, refer to chapter 20 in Vaughan et al., (2010) before proceeding.

The free (open source) software **Audacity** can be used to record and analyze animal sounds (Figure 12.1; http://audacity.sourceforge.net/). Another free sound analysis program is **Raven Lite** (*Raven Pro* is a full feature commercial version) produced by Cornell University Lab of Ornithology (Figure 12.1; http://www.birds.cornell.edu/brp/raven/RavenVersions.html#RavenLite). The discussion that follows will use *Audacity*. *Audacity* produces three types of graphs commonly used to visualize sounds. Oscillograms or waveforms display sound intensity fluctuations over the time (seconds). The x-axis represents time while sound volume (amplitude) is reflected in the height of the spikes on the y-axis (Figure 12.2). This is the default graphical display in *Audacity*.

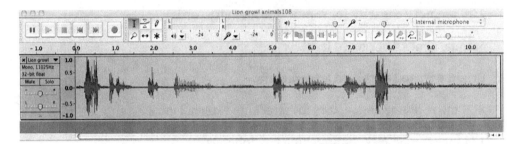

Figure 12.1 *Audacity sound analysis software is a free and open source product. Raven Lite is also free from Cornell University Lab of Ornithology. Both are available for Mac OS X, Microsoft Windows, GNU/ Linux, and other operating systems.*

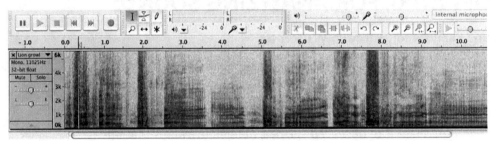

Figure 12.2 *A waveform (oscillogram) display of a lion growl in Audacity.*

The second type of display is called a sonogram (Figure 12.3). Sonograms include information on the frequency (pitch) of the sound. The x-axis again represents time in seconds and the y-axis displays frequency, with low-frequency sound near the bottom. The frequency of the lion growl in Figure 12.3 ranges from 80 Hz to 5,500Hz (or 5.5kHz). In a sonogram the amplitude (volume) is illustrated by the intensity of the colors. In black and white, the highest intensity sounds are black and the lower amplitude sounds in lighter shades of gray. In a color sonogram, amplitude variations are displayed in different colors (set by the user).

Figure 12.3 *A sonogram of a lion growl.*

The third type of graphical display is called a power spectrogram (Figure 12.4). Power spectrograms graphs frequency versus amplitude in decibels (dB) summed over the length of the sound selected (or over the entire song). Frequency is displayed on the x-axis and amplitude (volume in dB) on the y-axis. These plots are generated using fast fourier transformation (FFT) algorithms.

It shows the sound energy present at each frequency within the sound file. For this lion growl, the highest amplitude (loudest) sounds occur at frequencies around 390 Hz. This is not surprising given that lions have a very deep, throaty growl. For animal sounds, the most useful display is the frequency-time display, with intensity coded in greyscale or a color scale. This kind of analysis is usually called sonogram or spectrogram (Figure 12.3). These graphic displays of audio information allow researchers to compare signals from different species, different individuals, or the same individual in different behavioral contexts. Sonograms (or spectrograms) may show infrasounds, like those emitted by elephants, as well as ultrasounds emitted by echolocating bats (Figure 12.5). Spectrograms may reveal features, like fast frequency or amplitude modulations we can't hear even if they lie within our hearing frequency limits (30 Hz - 16 kHz). Spectrograms are widely used to show the features of animal calls or vocalizations.

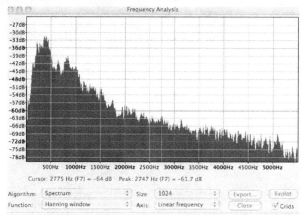

Figure 12.4   Power spectrogram for a lion growl.

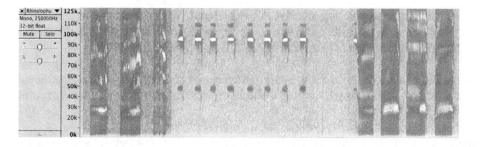

Figure 12.5   A sonogram of a Rhinolophus blasii *bat. Note the frequency range extends up to 125 kHz or well above the human hearing range.*

INTERPRETING A SONOGRAM

A tone with a constant frequency (such as a single note on a piano) will appear as an horizontal line with its vertical position depending on the frequency of the tone. A tone with increasing or decreasing frequency will appear as an inclining or declining line (Figure 12.6). For example, in the *Rhinolophus* signal on the left in Figure 12.6, the frequency is held constant at about 95 kHz for 0.04 seconds (40 milliseconds); this is referred to as a constant frequency (CF) echolocation signal. In contrast, the signal from *Nyctalus* on the right in Figure 12.6 shows echolocation pulses where the signal sweeps downward from a high of roughly 50 kHz to a low of 20kHz over 0.02 seconds (20 milliseconds). *Nyctalus* also displays a signal that has multiple harmonics. The lowest tone is called the fundamental frequency and tones that are multiples of this are called harmonics. For a detailed discussion of bat echolocation signals, readers are referred to chapter 20 in Vaughan et al., 2010).

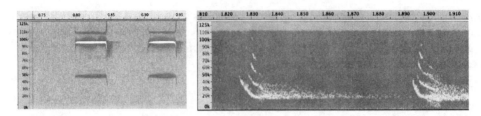

Figure 12.6 Two contrasting bat echolocation signals. (Left) a constant frequency (CF) signal from a bat in the genus Rhinolophus. *(Right) a frequency modulated (FM) signal from a bat in the genus* Nyctalus.

EXERCISES

EXERCISE 1: FIELD RECORDING

In this exercise you will record animal sounds in the field for later analysis. Your instructor will provide each group of students with a set of recording equipment and review the operation of the specific equipment provided.

1) Assemble the entire recording system and check the equipment in advance. Make a test recording (speak into the microphone) to make sure everything is connected properly.

2) Walk slowly down a trail near where you expect (from previous experience) you will encounter squirrels (or another mammal).

3) Try to avoid recording on windy days. Use your body to shield a shotgun mic from the wind. Wind speed is much less near the ground, so kneeling or hold the microphone as close to the ground will help reduce wind noise. If you are using a parabolic mic, holding it so that the wind is coming from the back of the dish.

4) Get as close to the squirrel as you can without disturbing it. Halving the distance to the animal doubles the signal level reaching the microphone.

5) Point the microphone directly at the vocalizing animal (try to avoid obstructions between the mic and the animal).

6) Set the record level so the loudest sounds register just into the red region (color may vary on different devices) of the VU meter (peak meter).

7) Record for at least one minute, or longer if the animal allows.

8) Minimize your body movements and other sounds.

9) Announce basic data at the end of each recording.

10) Listen through headphones to recordings you have just made to ensure it is recorded correctly.

11) Take turns making recordings so that every student in your group has several audio records.

12) Each time you record a new sound, record the information on your data sheet (see Appendix). A student who is not recording should fill out the data sheet.

13) Review and organize your field recording at the end of each day. The goal is to record as many vocalizations from as many animals as possible. These recordings may be used in exercises 2 and 3.

EXERCISE 2: SOUND ANALYSIS USING *AUDACITY*

In this exercise you will use *Audacity* software to analyze several audio recordings of mammalian sounds. Your instructor will provide you with access to the sound files you will analyze as part of this exercise. Additional sounds can be found at The Macaulay Library at the Cornell Lab of Ornithology -

http://macaulaylibrary.org/browse/scientific/10520807

or Mammal sounds on the Net -

http://thryomanes.tripod.com/MammalSounds.html

1. Download a lion growl from the *Mammal Sounds on the Net* webpage and then launch *Audacity*.

2. Click on *File | Open* and locate the audio file called "Lion growls" on your computer. These are .wav files. The audio file will be opened in *Audacity* (Figure 12.7).

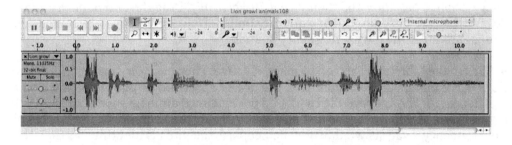

Figure 12.7   A waveform of a lion growling displayed in Audacity.

3. The "Waveform view" is the default view in *Audacity*. It shows the amplitude (loudness) of the sound wave on the vertical axis and time on the horizontal axis. Each of the "notes" shows up as a separate set of peaks. Play the lion growling audio file by clicking on the green *Play* button at the top left of the *Audacity* screen. A vertical line traces across the waveform as the sound file is played.

4. To the left of the waveform is an information box that lists the name of the file at the top (Lion growl) and details about the recording below (Mono, 11025Hz 32-bit, etc). Click on the small *black triangle* next to the name of the audio file. A drop down menu appears; Click on *Spectrum* from this list. *Audacity* now displays the same file

as a sonogram with the frequency (pitch) of the sound on the vertical axis in kilohertz and time in seconds on the horizontal axis. Loudness is indicated by colors.

5. Click *Play* to listen to this sound again, and you should see the pitch (frequency) rising and falling during some of the notes. Focus on the small segment at the end of the sonogram where the lion is producing a deep-throated "purring" sound.

6. Drag the mouse over the section from roughly 9.5 seconds to 10.5 seconds. Click on *View | Zoom to Selection*. Then click *View | Fit Vertically* (Figure 12.8). Click *Play* again to see how this set of notes sounds.

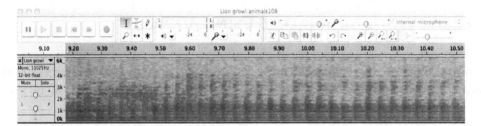

Figure 12.8 A close up of the last second and a half of the lion growling audio file.

7. In the menu bar at the top of the page, click on *Analyze* and select *Plot Spectrum*. The Frequency Analysis window shows the "spectrum" of the sound - a plot of the amplitude (loudness) of the sound wave on the vertical axis and frequency on the horizontal axis. What frequencies are the loudest? Quietest? How do you think this would compare with other large cats? To find out, Click on *File | Open* in the top menu bar and open the leopard growl audio file.

8. Play the leopard file (also downloaded from *Mammal Sounds on the Net*) and select a 1 second portion in the growl phase (e.g. form 1.0 to 2.0 seconds). Click on *View | Zoom to Selection* and then *View | Fit Vertically*. Drag the two *Audacity* window so they are lined up. Do they appear similar? How would you determine which animal produces the lowest frequency sounds?

9. Click in the Lion growl screen to make it active. Then click on *Analyze | Plot Spectrum* in the menu bar. You should see the power spectrum displayed for the Lion file. Repeat for the Leopard file to plot its power spectrum. The spectrum shows the highest peak at 905 Hz for the lion and at 167 Hz for the Leopard. At the bottom of the Frequency Analysis window for the Leopard, the pitch is identified as G#3, that is, G-sharp in octave 3. What is the pitch of the highest peak for the Lion?

10. Open the Cougar file. Select and zoom in on the section of the recording between 3.8 and 4.1 seconds (as described above in parts 5 and 6. Create a power spectrum of this segment of the Cougar file. How does it compare to a comparable section of the Lion file (you will need to locate and re-analyze the Lion file focusing in on the segment between 7.6 and 7.9 seconds).

11. Next you will look at a very different type of sound. In *Audacity*, open the file named "*Myotis daubentonii.*" (Figure 12.9). How long is this sound clip? It appears that there are elements of the sound file that are greater than the 20 kHz displayed by default.

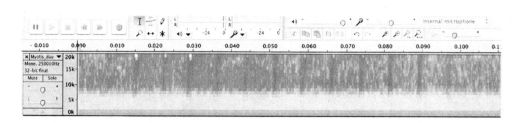

Figure 12.9 A sonogram of Myotis daubentonii *as displayed in Audacity over the default human hearing range of 0kHz to 20kHz.*

12. Click *Play* to listen to the sound. What do you hear? Probably nothing. Why is that? Can we use *Audacity* to generate an audible sequence for us?

13. Place the cursor over the y-axis and a small magnifying glass icon appears. Click on the *20 k* near the top if the y-axis. Notice that there are indeed sounds above 20 kHz. Repeat clicking on the upper end of the y-axis until you reach about 100 kHz. You want to display the entire sound on the screen at one time.

14. Hold down the *Shift* key (the magnifying glass has a minus symbol inside) and click just below the 100 k on the y-axis. This shrinks the y-axis to fit inside the screen. Stop when you have the y-axis running from 0k to 100k (Figure 12.10). Now you can see the entire range of sound.

Figure 12.10 A sonogram of Myotis daubentonii *showing the full spectrum of sound.*

15. *Play* the sound. Can you hear it?

16. Click on *Effect | Change Speed* and when the pop up screen appears set the slider to -50.0 (slows it down by 50%). Click *OK*. Now *Play* the sound. should be able to hear the high frequency echolocation signals now.

17. Plot the power spectrum for this call. What is the loudest frequency? What is the lowest frequency? Is the lowest frequency within the human hearing range? Why don't we hear it? (Hint: look at the dB). Now that you have seen how to analyze sounds using *Audacity*, you should search for a set of 3-4 mammal sounds on the web.

18. Create a set of hypotheses and predictions about these sounds. It may help to choose sounds from related species. For example, you might choose to compare large felids (lion and tiger) with smaller cats (bobcat or house cat), or you might compare the sounds made by various species of whale or primates. Alternatively

you could analyze several different calls from the same species (e.g. territorial call versus alarm calls).

19. Download those wav or mp3 files to your computer (from the sites listed previously), and analyze them in *Audacity*.

20. Write a report, in a format specified by you instructor, that describes you hypotheses and predictions and the results of your analysis.

EXERCISE 3: PLAYBACK EXPERIMENTS USING ALARM CALLS

In this exercise you will conduct a playback experiment on squirrels, chipmunks, or some other locally abundant species (see Greene and Meagher, 1998 for a similar study). Your instructor will provide you with a set of audio tracks for playback in the field. You will record the squirrel's behavior in response to each playback. A variety of calls can be used. For example, your instructor may give you:

- an alarm call from the squirrel species you are observing (but not from the same individual),

- the call of an aerial predator (bird of prey) native to your region (e.g. red-tailed hawk), and

- a segment of "white noise" that serves as the control sound.

Note: Your field site should be a location where squirrels are abundant, easily visible, and relatively tolerant of human activity. It may help to set up feeding stations baited daily with sunflower seeds for several weeks in advance of the study to attract squirrels to the area you've chosen.

One trial consists of playing back one track. For each trial you will record:

- travel time to the feeding station,

- if travel was direct, with no stops on the way to the feeding station, or periodic, with short stops to scan for intruders.,

- handling time at the feeding station (total time spent handling, eating, or manipulating seeds),

- frequency and duration of vigilant behaviors (looking up, scanning area, etc.),

- residence time (total time at station),

- distance a subject animal ran to cover (after hearing the playback stimulus),

- recovery time (time it took to re-emerge and resume activity), and

· number of seeds removed (count remaining seeds at the station after the trial).

PROCEDURES

1) Locate the site where feeding stations have been in use and set up your observation post about 10 meters away.

2) Set up the playback speaker (self-powered speakers that run on batteries) 3-4 meters from the feeding station and camouflage it as best you can.

3) Add exactly 60 sunflower seeds to the feeding station (remove any remaining seeds from previous feedings).

4) Return to your observation post and allow time for the subject to resume normal activity.

5) Begin trial #1 by recording the pre-playback behavior of your subject animal for several minutes.

6) After your subject has been at the feeding station for 15-20 seconds, broadcast the first playback track (either a control sound or alarm call).

7) Simultaneously note the animal's behaviors during the playback (video taping may be helpful).

8) After the playback, continue recording behavior every 30 seconds for the next five minutes (i.e. the post-playback segment). This is the end of trail #1. Wait 30 minutes before proceeding to trial #2.

9) Count the seeds remaining at the feeding station and replace with 60 new seeds to begin trail #2.

10) Wait patiently for the subject to return to the feeding station and resume activity.

11) Play the second track while recording the subject's response as before. This ends trail #2. Wait 30 minutes before proceeding to trial #3.

12) Count the seeds remaining at the feeding station and replace with 60 new seeds to begin trail #3.

13) Play the third track while recording the animal's behavior as before. Thus, each subject receives all three playback treatments on the same day.

14) Collect your equipment and move to another study site (if instructed to do so) and repeat the process using a different subject animal.

After the experiment has been completed, the instructor will tell each group the type of call on each track. Because the observers don't know ahead of time what call is given at any time, this is called a blind experiment. Label each set of pre, during, and post behaviors with the track number and call type for your group. Groups will then pool their data for analysis. Your instructor will describe how the data should be graphed, analyzed statistically, and reported.

BIBLIOGRAPHY

Ahlén, I. & H. Baagøe. (1999) Use of ultrasound detectors for bat studies in Europe – experiences from field identification, surveys and monitoring. *Acta Chiropterologica*, 1:137-150.

Goodall, J. (1986) *Chimpanzees of Gombe*. Harvard University Press, Cambridge, MA.

Greene, E. and T. Meagher. (1998) Red squirrels, *Tamiasciurus hudsonicus*, produce predator-class specific alarm calls. *Animal Behavior*, 55:511-518.

McComb, K., Reby, D.,Baker, L., Moss, C. and S. Sayialel. (2003) Long-distance communication of acoustic cues to social identity in African elephants. *Animal Behaviour*, 65:317-329.

Payne, K. B., Langbauer, Jr., W. R. and E. M. Thomas. (1986) Infrasonic calls of the Asian elephant (*Elephas maximus*). *Behavioral Ecology and Sociobiology*, 18:297–301.

Poole, J. H., Payne, K., Langbauer, Jr., W. R. and C. J. Moss. (1988) The social contexts of some very low frequency calls of African elephants. *Behavioral Ecology and Sociobiology*, 22:385–392.

Randall, J. A. (1989) Individual footdrumming signatures in bannertailed kangaroo rats, *Dipodomys spectabilis*. *Animal Behavior*, 38:620–630.

Randall, J. A. (1997) Species-specific footdrumming in kangaroo rats: *Dipodomys ingens, D. deserti, D. spectabilis*. *Animal Behavior*, 54:1167–1175.

Randall, J. A. and M. D. Matocq. (1997) Why do kangaroo rats (*Dipodomys spectabilis*) footdrum at snakes? *Behavioral Ecology*, 8:404–413.

Vaughan, T., Ryan, J. M. and N. Czaplewski. (2010) *Mammalogy*, 5th edition, Jones & Bartlett Publishers, Sudbury, MA.

Weilgart, L. and H. Whitehead. (1997) Group-specific dialects and geographical variation in coda repertoire in South Pacific sperm whales. *Behavioral Ecology and Sociobiology*, 40:277–285.

APPENDIX

Date:	Location:		
GPS Coordinates:	Lat.		Long.
Weather			
Recorder Settings:			

Audio file #	Counter Range	Time	Species	Behavior notes

13

Quantifying Mammalian Behavior

TIME REQUIRED

One 3-4 hour lab period. Additional time may be needed.

LEVEL OF DIFFICULTY

Basic

LEARNING OBJECTIVES

Understand how to categorize behaviors.

Learn how to construct an ethogram.

Learn different types of sampling in mammalian behavior.

Practice behavioral observation techniques in the field.

Understand how to avoid common problems.

EQUIPMENT REQUIRED

Binoculars

Field notebook

Data sheets and clipboard

Digital stop watch

BACKGROUND

By now you are familiar with the scientific method and hypothesis testing, where observations lead to questions, which in turn lead to specific hypotheses. Hypotheses are nothing more than a reformulation of the question into a formal, testable statement. For example, suppose you observed a group of foraging elk and wondered why certain individuals seem to spend more time looking around (e.g. vigilance behavior) than other members of the herd. This is essentially your question - why do certain individuals spend more time being vigilant than others? To create a testable hypothesis, you need to restate this question more formally. You might state your hypothesis as "dominant individuals spend more time being vigilant that subordinate elk." Recall that the hypothesis is a statement (not a question) with one potential explanation for the observations. This hypothesis is testable because you can collect behavioral observations on dominance level and amount of time spent being vigilant for each individual. In essence your hypothesis gives you a plan for collecting additional data. At this point, it's back to the field to collect the data needed to test your hypothesis. If your new data falsify your original hypothesis, then you choose an alternative hypothesis that potentially explains your observations. For example, you might restate your hypothesis as "individuals foraging on the perimeter of the herd spend more time being vigilant than those at the center of the herd."

This process sounds simple, but in practice you need to know enough about the behavior of elk to determine dominant behaviors from submissive behaviors and vigilance from other types of behavior. The first step in studying the behavior of a species is to construct an ethogram. In its simplest form, an ethogram is a catalog of an animal's behaviors (Altmann, 1974). Constructing ethograms requires careful observations (notes) of each behavior. The result is a complete list of behaviors with a definition of each. Ideally, each behavior should be unambiguous; it should not be possible to assign a new observation to more than one category in your ethogram. The level of detail in your ethogram depends on the goals of the study and the hypothesis being tested. For example, suppose you observe a raccoon digging in the muck at the margin of a stream, extracting a clam, breaking open the clam's shell on a rock, and extracting and eating the meat. Each of these events might be a separate category (with its own definition) in a study testing hypotheses of foraging behavior. However, if the goal was to construct a time budget for raccoon daily activity, you might combine all of these behaviors into one category called "foraging."

AVOIDING COMMON PROBLEMS

Before you begin studying animal behavior, you should understand some common pitfalls of behavioral studies. Among the most important decisions is how many individuals to observe (e.g. to have a large enough sample size for statistical tests). It may be much easier to observe many behaviors on one individual, but this leads to pseudoreplication (treating the data as independent observations when they are in fact interdependent, Hurlbet, 1984). Observations need to be independent events. If you measure the same behavior multiple times in the same individual, the behaviors are not independent events (e.g. not true replicates). There are times when your study design may require you to take repeated measures on the same animal (e.g. behaviors before and after an environmental disturbance). In general, the best way to characterize an animal's natural behavior is to watch as many different individuals as possible, then summarize their "average" behavior. Ideally, you can tell individuals apart. When that is not possible, you should watch a large enough group that there is a low probability of watching the same individual twice.

Another problem that can crop up is observer bias. An observer who knows the hypothesis may unintentionally interpret behaviors as favoring the hypothesis (or ignore results that do not favor the hypothesis). Such bias can be avoided by conducting a "blind" study in which the observer(s) do not know the hypothesis or its predictions in advance. Of course this is not always practical and observers should always strive for complete, honest, and unbiased records of behavior.

Often two or more observers record data that will be pooled for analysis (e.g. pooling class results to yield larger sample sizes). In such cases it is very important to know that each observer is recording the same thing. For example, one observer might score a behavior differently than another observer. One way to prevent inter-observer variability is to properly train the observers beforehand and cross-check their results. Observers should also be randomly assigned to study groups. This is especially important if there are separate treatment and control groups. Having one observer record only treatment groups and another observe only control groups leads to invalid results. There are other issues to avoid as well, but these are the most common pitfalls in behavioral studies.

EXERCISES

EXERCISE 1: BUILDING AN ETHOGRAM

In this exercise you will construct an ethogram for a species chosen by your instructor (Squirrels are an obvious choice in many regions, because many species are diurnal and easily observed). In your fieldwork, you first must characterize the range of behaviors you see. Define behaviors in terms of the animal's actions, not by function. For example, if you see a squirrel bury an acorn, you would record it as "burying" and define it as digging and placing an acorn in the ground, and covering it. You would not record it as "storing food" because this implies a function or interpretation you did not observe directly- that the animal would return at some later point to retrieve the acorn. One of the hardest lessons of animal behavior is learning to avoid projection (either of adaptive function or anthropomorphism).

PHASE I: INITIAL OBSERVATIONS

· Find a suitable spot to observe you target animals. Make sure you are far enough away to avoid disturbing the animal's behavior, but close enough to see what is happening.

· Sit down, get comfortable, and begin to observe and record behaviors. Avoid excess movements or noise.

· In you field notebook, record all the different kinds of behavior you observe. Only after you have built a catalog of possible behaviors for your animal, will you be ready to quantify those behaviors in terms of frequency, duration, and context.

· Initially, you should note the context in which behaviors are performed. "Context" includes location (habitat type, proximity of major features of the habitat), other organisms (same or different species), physical conditions (time/temperature/light), previous behavior, individual characteristics (male/female, nutritional state, reproductive state, size) if known. One way to do this is to create a log of all the behaviors as they occur, including what is the next behavior in the sequence and, if possible, how much time each behavior lasted.

PHASE II: BUILDING THE ETHOGRAM

Now that you have a reasonably complete set of field observations, it it time to begin organizing the behavioral catalog or ethogram. The ethogram is a list of behaviors and an explicit definition for each. The catalog allows you to organize your data sheets so that you can quantify behavior. For example, some grey squirrel (*Sciurus carolinensis*) behaviors are shown in Table 13.1.

Table 13.1 Sample behavior catalog for a gray squirrel.

Behavior	Definition
running	bounding movement over the ground
climbing up	vertical movement away from ground
climbing down	vertical movement towards the ground

Once you have the basic catalog with objective definitions, you can organize your catalog according to behavior types. In the above list, "running", "climbing up", and "climbing down" might fall under a category called "movement", whereas "chirp" and "chatter" might fall under "vocalization". Such a hierarchy will let you organize your data sheets and your analyses more easily. You should also assign each behavior category a unique, easy to recall, letter code (e.g. R for running and D for climbing down).

EXERCISE 2: SAMPLING BEHAVIORS

Once you have completed your catalog of behaviors, you are ready to quantify how your animal spends its time. You are expected to observe your study animal for at least three hours. You should aim for at least four individuals (see above) to be observed in this time period. You should devote an approximately equal amount of time to each individual you observe. You should be careful about when you observe them. It is best if you can observe each individual multiple times (i.e, three twenty minute observations on the same individual is better than a single one-hour observation on that individual). You should distribute your observation periods through the day so that you can get a generally representative sample of behavioral patterns. Alternatively, you should restrict your observations to particular time(s) of the day and distribute your observations evenly within those constraints.

So far what you have done is called ad lib sampling, but there are a number of methods for sampling behaviors:

Ad lib sampling: behaviors are recorded in the order they occur. Bias can creep into ad lib sampling if the observer tends to notice and record certain behaviors over others.

Scan sampling: observer scan the entire group of animals and records what each animal is doing. This approach works well for smaller groups. It is used to construct time budgets and/or for studying individual variation. A further advantage is that it is unbiased because every animal is sampled each time. Typically, scans are taken every 30 seconds, every minute, or at some other predetermined interval.

Focal animal sampling: a single individual is selected and followed. Its behaviors are recorded for a standard time (or until the animal moves out of sight). If a focal individual moves out of your view, then you start a new sequence of observations on a new focal individual. Selecting the focal animal can be systematic (e.g. follow only young animals) or randomized (select a random number from a table, then follow the nth individual encountered). The main disadvantage of focal animal sampling is that you can only record data from one animal at a time, so it will take longer to acquire an appropriate sample size.

Organized data sheets are critical for behavioral studies, because they allow you to record data quickly and efficiently during the observation period, and to tabulate the results accurately afterward. A well written catalog of behaviors will allow you to create a useful data sheet, which in turn simplifies the quantification phase of the study. A very simple example data sheet is shown in Table13.2.

· Create a data sheet from your ethogram.

· Return to the field and preform a scan sample every 1 minute for 1 hour.

· Fill in your data sheet by recording how many individuals are exhibiting each behavior at the time of the scan.

· Preform a focal animal sample by arbitrarily choosing one animal in the group and recording its behaviors on a new data sheet. Record each behavior in turn and the time that behavior lasts.

Table 13.2   Sample data sheet for a study of aggressive behaviors.

Time	Behavior (code)					
1 min intervals	Chase	Fight	Attack	Bite	Submit	Retreat
14:00						
14:01						
14:02						

EXERCISE 3: CREATING A TIME BUDGET

In exercises 1 and 2 you learned how to catalog and sample behaviors. The more interesting questions in animal behavior require us to detect relationships or patterns among the various behaviors animals preform. One way to detect patterns is to record the time and sequence of each behavior. The objective is to determine the probability that one behavior will be followed by (or preceded by) another behavior. Such sequence diagrams and time budgets can be derived from focal studies. In this exercise you will create a time budget analysis.

A time budget indicates the amount or proportion of time that animals spend in different behaviors, or in performing different classes of behavior. To construct a time budget, you must add the time spent on each behavior for each subject animal. Then calculate the proportion of time spent on particular behaviors, out of the total observation time. When you are able to observe several different individuals, then each individual can be a "replicate" or independent estimate of time allocation of the behavior of interest. You summarize your data by doing a time budget for each individual, then calculating mean values for each behavioral measure of interest (Table 13.3).

CREATING A TIME BUDGET:

1) Calculate the mean time spent for each behavior for all animals in the study and convert this to percent.

2) Create a table similar to Table 13.3 to summarize your results and write up a short results section in journal format.

3) List at least one hypothesis that you could test using the data from your time budget.

Table 13.3   *A partial time budget for large mammalian herbivores on the National Bison Range, Montana. Data are from Belovsky and Slade, 1986.*

Species	% Inactive Behaviors		
	Lying	Standing	Ruminating
Antilocapra americana	33 ± 15	14 ± 7	14 ± 3
Ovis canadensis	38 ± 20	6 ± 6	15 ± 15
Odocoileus virginianus	50 ± 60	5 ± 6	11 ± 4
Odocoileus hemionus	30 ± 27	11 ± 14	12 ± 13
Cervus elaphus	48 ± 33	7 ± 9	13 ± 20
Bison bison	40 ± 31	14 ± 5	20 ± 3

EXERCISE 4: CREATING A TRANSITION DIAGRAM

The second method for analyzing behavior patterns is to derived a transition matrix. A transition matrix expresses the probability that one behavior is followed by another. For example, we might expect that digging a hole is followed by burying an acorn with a probability of 0.85 (85% of the time), but digging a hole is followed by chasing another squirrel with a probability of only 0.15. A transition matrix links "chains" of behaviors and is a technique to look at relationships between different behaviors. Creating a transition matrix requires knowing the sequence in which each behavior was preformed.

In a transition matrix, behaviors are listed in rows and columns in a table. The "cells" of the table are the relative frequency that a reference (row) behavior is followed by the next (column) behavior. A simple example of a transition matrix is shown in Table 13.4.

Table 13.4   An example of the raw data for aggressive behaviors used to create a transition matrix.

Beginning Behavior	Followed by Behavior						Totals
	Chase	Fight	Attach	Bite	Submit	Retreat	
Chase	0	21	8	3	67	54	153
Fight	31	0	2	45	56	87	221
Attack	23	69	0	15	45	91	243
Bite	1	21	0	0	78	23	123
Submit	0	1	6	0	0	0	7
Retreat	0	0	0	0	5	0	5
Totals							752

Suppose your first sequence of behaviors was a chase followed by an attack, followed by a bite. You would score the chase-attack sequence as a 1 in row one (chase), column 3 (attack) and a 1 in row three (attack), column four (bite). You continue tallying each sequence until all the aggressive behaviors have been scored and added to your table. The result is a matrix that shows the number of times each behavior listed in column one on the left (Beginning Behavior) is followed by each behavior listed in the row across the top (Followed by Behavior). Thus, it gives the number of times there was a transition from one type of behavior to another (a behavior can follow itself if two of the same behaviors occurred in sequence).

You next calculate the percentage of times that a particular action pattern follows another (given the first action pattern). For example, in Table 13.4, 21 chases were followed by fights out of a total of 153 chases observed. The transition probability is $21/153 = 0.137 = 13.7\%$. Table 13.5 illustrates the transition percentages from the raw data in Table 13.4.

Table 13.5   Transition percentages for the aggressive behavior data presented in Table 13.4.

Beginning Behavior	Followed by Behavior						Totals
	Chase	Fight	Attach	Bite	Submit	Retreat	
Chase	0%	14%	5%	12%	44%	35%	100%
Fight	14%	0%	1%	20%	25%	39%	100%
Attack	9%	28%	0%	6%	19%	37%	100%
Bite	1%	17%	0%	0%	63%	19%	100%
Submit	0%	14%	86%	0%	0%	0%	100%
Retreat	0%	0%	0%	0%	100%	0%	100%
Totals							100%

Please note that transition percentages across a row must sum to approximately 100, but individual columns do not need to add to 100%. Can you explain why this would be the case?

CREATING A TRANSITION DIAGRAM

A transition diagram (aka kinematic diagram) is simply a flow chart of the behaviors. To make one:

- Start by drawing a circle or box and place the abbreviations for one behavior in the box. You can begin with any behavior. You may want to add sample sizes in parentheses (from the raw data matrix, right column totals) into the box with each behavior (Figure 13.1).

- Draw arrows to the other behavior circles that followed this behavior (Figure 13.1).

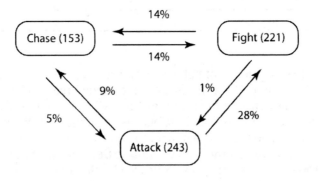

Figure 13.1   Preliminary transition diagram for the data in Table 13.5.

- Next to each arrow place the percentage from the transition matrix. Be sure to note the direction of the arrow.

Transition diagrams allow you to see quickly which behaviors follow which others and gives the frequency with which these sequences of behavior occur. By examining transitions over different individuals or in different conditions, you may learn about "complexes" of coordinated behaviors. Also, you can ask questions about how different conditions influence the flow of behaviors through time.

EXERCISE 5: CREATING A DOMINANCE HIERARCHY

A dominance hierarchy analysis is a record of winners and losers in encounters that is used to determine dominance ranks among individuals. For example, suppose you were interested in establishing the dominance hierarchy for a small herd of elephants from Amboseli National Park, Kenya. One such family group hierarchy is show in Table 13.6 from Archie et al., 2006.

In Table 13.6, the letters represent one individual in the CB family unit, and the intersection of rows (aggressor) and columns (loser) is the number of interactions won by aggressor (rows). For example, elephant A won 6 times against elephant B and 7 times against elephant D, but only once against elephant C. Elephant A is clearly the dominant animal in the group. Can you rank dominance of these elephants based on information in this table?

Table 13.6   Dominance matrix for the CB family unit of elephants from Amboseli National Park, Kenya (Data from Archie et al., 2006).

	A	B	C	D	E	F
A	-	6	1	7	6	2
B		-	1	3	6	1
C			-	2	1	1
D	3			-	5	2
E		1			-	
F						-

Dominance hierarchies often are linear, showing a specific pecking order within the group. In the elephant example in Table 13.6, elephant A dominates everyone else, elephant B dominates everyone but elephant A, and so on. In other cases, the hierarchy may not be quite so linear. One way to test for linearity is to use Landau's H.

$$H = \left(\frac{12}{n^3 - n} \right) \sum \left[v_a - \frac{(n-1)}{2} \right]^2$$

where,
   n is the number of animals in the group

   v_a is the number of animals dominated by animal a (e.g. the number of individuals that individual outranks)

   Landau's H ranges from 0 to 1 with values near 1 being more linear hierarchies and a 0 indicates no hierarchy.

   1) Calculate Landau's H for the elephant group in Table 13.6.

   2) Construct a dominance hierarchy matrix from data you have collected and determine if your animals exhibit a dominance hierarchy.

EXERCISE 6: DOMINANCE HIERARCHY ANALYSIS

   You will now use a Java applet (an online software tool written in the Java Programing language) to examine dominance hierarchies in an example elephant data set and in the data you collect on your subject species in the field.

   1. Point you web browser at
          http://caspar.bgsu.edu/~software/Java/1Hierarchy.html

2. Assuming your browser supports Java, you should see a description of the Java applet followed by a series of three contingency tables at the bottom of the page (Figure 13.2).

3. You can set the size of the data matrix using the drop-down buttons at the very bottom of the web page. In your elephant example from Table 13.6, we have a 6 x 6 matrix. Set the applet's matrix size to 6 x 6, using this button at the bottom of the page.

4. Click on the first available cell (e.g. cell 1B) in the applet matrix - it turns light blue when it is available for data entry. Use the data from Table 13.6 to fill in the upper right part of the applet table (Figure 13.3).

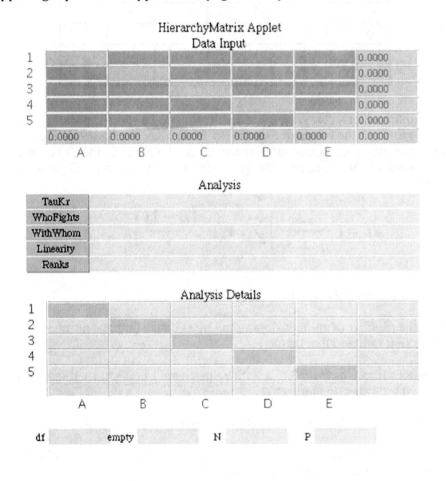

Figure 13.2   The Java Data Grinders applet for dominance hierarchy analysis.

Note: you must click on each cell, type in a number, and then click on the next cell - if you make a mistake you have to reload the page (there is no reset button to clear the table).

HierarchyMatrix Applet
Data Input

	A	B	C	D	E	F	
1		6.0000	1.0000	7.0000	6.0000	2.0000	22.0000
2			1.0000	3.0000	6.0000	1.0000	11.0000
3				2.0000	1.0000	1.0000	4.0000
4	3.0000				5.0000	2.0000	10.0000
5		1					1.0000
6							0.0000
	3.0000	7.0000	2.0000	12.0000	18.0000	6.0000	48.0000

Figure 13.3   Example of the dominance data entered into the applet.

5. With the data entered you can now examine the output of several measures. First, click on the *Linearity* button on the left side of the analysis table (Figure 13.4). You will see your value for Landau's H listed here.

· Does this value correspond well with the value you calculated by hand?

· Does this value suggest that the dominance hierarchy for this family unit is linear?

Analysis

TauKr	
WhoFights	
WithWhom	
Linearity	Landau h: 0.9714; circ. triads: 0; Appleby K: 1.0000
Ranks	

Figure 13.4   Linearity analysis for the elephant data.

6. Next click on the *Ranks* button in the analysis table. This will display the calculations for several ranking methods (Figure 13.5). The first two measures are the ordinal ranks and dominance index (from Theraulaz et al., 1992). These both indicate that elephant A in our example is dominant to all others in the group, and that elephant B is dominant to all others except elephant A. The third column shows the dominance activity index (DAI) calculated as follows (from Bartos, 1986):

$$DAI = \log\left\{\frac{(P+0.1)^2}{N+0.1}\right\} + 1$$

Analysis

TauKr	
WhoFights	
WithWhom	
Linearity	
Ranks	Boyd&Silk: -1 iterations; BBS: -1 iterations

Analysis Details

	A	B	C	D	E	F
1	1.0000	0.8696	5.8700	◆	-1.1030	
2	2.0000	0.5882	3.6650	◆	-0.1565	
3	3.0000	0.6000	2.5209	◆	-0.3129	
4	4.0000	0.4000	2.6905	◆	0.1449	
5	5.0000	0.0526	-1.7053	-201.8165	1.4051	
6						
	ordinal	dom. index	DAI	Boyd&Silk	BBS	

Figure 13.5   Analysis of the ranks of the elephant family unit using five methods.

The final two columns display ranks based on methods by Boyd and Silk (1983) and the Batchelder-Bershad-Simpson (BBS) method (Jameson et al., 1999). The Boyd & Silk method, will not work properly if we lack information about inter-actions between some pairs of individuals or if the group includes individuals who are never dominated by others. The BBS method, however, does provide stable ranks even if both of these conditions are not met. In the BBS method an animal's rank depends on (1) its proportion of wins, (2) its proportion of losses, and (3) the scores of the others it encountered (Jameson et al., 1999). In this case low scores represent higher ranks. The BBS model uses the following equation to provide initial estimates of rank:

$$s(a_i) = \left[\frac{a(2W_i - N_i)}{2N_i} \right]$$

where,

is a constant

$$a = \sqrt{2\pi} = 2.50663$$

W_i is the number of encounters in which animal a_i won, and

N_i is the total number of encounters for animal a_i.

Each animal is assigned an initial score $s(a_i)$ based on the proportion of in-teractions it won. Then, a second equation is used rescale (recursively) all animals repeatedly until their scores do not change. The second equation is:

$$s(a_i) = \left[\frac{2W_i - L_i}{2N_i} \right].$$

where,

L_i is the number of encounters animal a_i lost, and

Q_i is the mean score of animals that a_i encountered.

7. Use the data you gathered from your field observations and rerun the domi-nance hierarchy applet. Write a short summary of your results1.

---

[1]This example is based on a collection of public domain Java Applets for the analysis of be-havioral data (available on the Internet at: http://caspar.bgsu.edu/~software/Java/).

BIBLIOGRAPHY

Altmann, J. (1974) Observational study of behavior - sampling methods. *Behaviour*, 49:227-267.

Archie, E. A., Morrison, T. A., Foley, C. A. H., Moss, C. J., and S. C. Alberts. (2006) Dominance rank relationships among wild female African elephants, *Loxodonta africana. Animal Behavior*, 71:117-127.

Bartos, L. (1986) Dominance and aggression in various sized groups of red deer stags. *Aggressive Behavior*, 12:175-182.

Belovsky, G. E. and J. B. Slade. (1986) Time budgets of grassland herbivores: body size similarities. *Oecolgia*, 70:53-62.

Boyd, R. and J. B. Silk (1983) A method for assigning cardinal dominance ranks. *Animal Behavior*, 31:45-58.

Hurlbert, Stuart H. (1984) Pseudoreplication and the design of ecological field experiments. *Ecological Monographs.* 54:187–211.

Jameson, K. A., Appleby, M. C., and L. C. Freeman. (1999) Finding an appropriate order for a hierarchy based on probabilistic dominance. *Animal Behavior*, 57:991-998.

Theraulaz, G., Gervet, J., Thon, B., Pratte, M. and S. Semenoff-Tian-Chansky. (1992) The dynamics of colony organization in the primitively eusocial wasp *Polistes dominulus* (Christ). *Ethology*, 91, 177-202.

Vaughan, T., Ryan, J. M., and N. Czaplewski. (2010) *Mammalogy*, 5th edition, Jones & Bartlett Publishers, Sudbury, MA.

14

Optimal Foraging Behavior

TIME REQUIRED

One 3-4 hour lab period for each of the exercises

LEVEL OF DIFFICULTY

Moderate

LEARNING OBJECTIVES

Understand the basics of optimal foraging theory

Understand how energy and profitability are calculated

Understand how handling time, search time, and other factors shape foraging decisions

Learn how animals make decisions about which patch to forage in

Apply optimal foraging theory to squirrel foraging in the field

EQUIPMENT REQUIRED

Clipboards

Data sheets

Peanuts (shelled and unshelled) and sunflower seeds

Two foot square plywood trays (painted white)

Tape measures

Stop watches

Binoculars

BACKGROUND

FORAGING COSTS AND BENEFITS

All mammals are predators at some level. Wolves are predators that eat elk, deer, and moose. Kangaroo rats are capturing helpless seeds. Whether they are carnivores or herbivores, all mammals must search for and capture food. In doing so they must make choices about what type of food to search for, how long to spend searching, and where to search. Optimal foraging theory attempts to understand why certain foraging behaviors are adaptive (Krebs, 1978). The basic prediction is that animals will forage in a way that maximizes net energy intake per unit time (Kolmes et al., 1997). Doing so ensures maximum fitness in terms of survival and overall reproductive success.

Net energy intake rate is determined by several factors, including:

- energy spent searching for food,

- energy value of the prey item,

- energy spent handling or capturing prey,

- the abundance of the food items in the animal's environment, and

- the distribution of the food items in the environment.

Thus, there are costs and benefits associated with foraging (Kolmes et al., 1997; Charnov, 1976). Over time, natural selection shapes behaviors that maximize fitness. It is therefore assumed that an animal's foraging behaviors are an important component contributing to its overall fitness.

Each food item has its own handling time, denoted by h. It is the amount of time it takes the predator to subdue, prepare, and consume a prey item once it has located the item. In addition, each prey item has a unique net caloric value (e.g. energy content) denoted by E. It is the amount of energy that can be extracted by one prey item via absorption across the intestines minus the energy spent to chew and digest the item. Thus, each prey item, whether it is a gazelle or a seed, has a unique profitability, denoted by P. Where P is the net caloric value of a prey item divided by the handling time (units are typically calories/ minute or Joules/second).

$$P = \frac{E}{h}$$

Virtually all mammals live in environments that contain a mixture or food items (e.g. prey species). Some prey are more profitable than others, and some prey are more abundant that others. Optimal foraging theory predicts that predators should select prey items in a way that the prey profitability and prey density (abundance) combine to yield the greatest net energy gain. In other words, if the most energy rich prey is rare, it may not pay to spend all the energy needed to search for a rare prey when other less profitable prey are easily at hand. Where multiple prey are available, what is the optimal foraging strategy for the predator?

From the predators point of view, prey density is really prey encounter rate. The denser the prey item, the more frequently the predator will encounter it. Encounter rate is denoted by λ, where the encounter rate for a prey of type i is λ_i = total encounters with prey type i divided by total search time. For example, a foraging otter might encounter 20 fish in a 10 minute interval yielding an encounter rate for fish of λ_{fish} = 2.0 fish per minute.

The encounter rates of each prey species are important because they influence the search times. To see why, consider a predator searching for food in an environment where prey are scarce. Assume it takes the predator t seconds to locate a particular type of prey. If we double the number of prey types searched for, and we assume equal prey densities for both prey species, then there should be twice as many encounters in the same time period. Stated differently, $t/2$ of search time is cut in half. If we now add a third prey type (of equal density), the encounter rate becomes $t/3$ and so on. Thus, search time is inversely proportional to the number of prey types available (Figure 14.1).

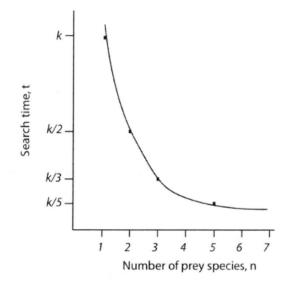

Figure 14.1   The inverse relationship between search time t, and the number of prey types, n, in a predator's diet.

Notice that when there are few prey species the graph is steep, but when more prey types are present it flattens out. What this means in practice is that a small increase in the number of prey species can dramatically reduce search costs for the predator. Thus, a predator can't be too fussy about which prey species it seeks or it may end up spending huge amounts of time searching.

Of course, so far we have assumed that each prey species is equally abundant. This is rarely the case in the real world. If a predator is in a good habitat where prey of all species are abundant, then search times will be small for each prey type. However, if prey are less abundant in poor habitats, then search times increase proportionally (Figure 14.2).

Recall that predators should try to maximize net caloric intake rate *(E/T)*. In this case, the model is concerned with maximizing energy, but it is possible to use other "currencies." For example, the energy maximizing model ignores the forager's need to avoid predators too. We might want a model that minimizes the time spent foraging (and therefore the risk of being preyed upon).

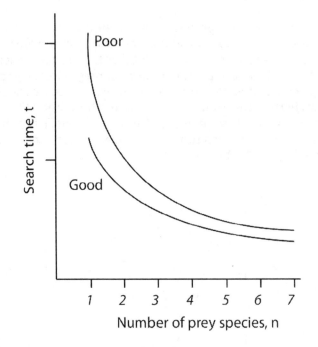

Figure 14.2 Search time t, and the number of prey types, n, in a good and a poor habitat. Notice that the steepness of the line for small numbers of prey (n) increases more sharply in the poor habitat.

A simple model of prey choice

If a predator specializes on only one prey type, then the net caloric intake rate would be

$$\frac{E}{T} = \frac{\lambda_1 E_1}{1 + \lambda_1 h_1}$$

where:
    λ_1 is the encounter rate of the prey,
    E_1 is the energy content in calories of the prey item, and
    h_1 is the handling time for that prey.

Suppose, however, we want to model how a forager would maximize its energy intake rate when given a choice of two different prey species. In this case, the encounter rates of the two prey are different (e.g. the prey have different densities) and we can denote them as λ_1 and λ_2. The net caloric values of the two prey species are denoted as E_1 and E_2, and their handling times are h_1 and h_2. To simplify, assume the predator is not selective (won't ignore one prey over the other). This means that in t minutes of searching, the predator will have encountered prey 1 at a rate of $\lambda_1 t$ and the second prey type at $\lambda_2 t$. The overall caloric intake rate is:

$$\frac{E}{T} = \frac{t\left(\lambda_1 E_1 + \lambda_2 E_2\right)}{t\left(1 + \lambda_1 h_1 + \lambda_2 h_2\right)} = \frac{\lambda_1 E_1 + \lambda_2 E_2}{1 + \lambda_1 h_1 + \lambda_2 h_2}$$

Similarly, if there are *n* different prey species in the predator's diet, then non-selective predation would give a net caloric intake rate of:

$$\frac{E}{T} = \frac{\lambda_1 E_1 + \lambda_2 E_2 + \ldots + \lambda_n E_n}{1 + \lambda_1 h_1 + \lambda_2 h_2 + \ldots + \lambda_n h_n}$$

The math looks complicated but is really quite simple (detailed examples and mathematical proofs can be found in Kolmes et al., (1997). These equations can be used to determine exactly which prey species should be included in the forager's diet. For example, if we know the profitabilities of prey species 1 is greater than that of prey type 2, e.g.

$$E_1/h_1 > E_2/h_2$$

and we want to know if it is more profitable to add prey 2 to the diet or simply feed only on prey 1, then we need to know which diet maximizes net caloric intake rate. In this case prey species 2 should be added to the forager's diet only if its profitability is greater than the net caloric intake rate for prey species 1 alone. In other words, if:

$$\frac{\lambda_1 E_1}{1 + \lambda_1 h_1} < \frac{E}{h}$$

Similarly, a third prey type should be added to a diet consisting of prey types 1 and 2 only if the intake rate of the first two prey items is less than for all three prey items. The equations above can be used to create a set of foraging rules. If there are *n* prey types in a possible diet and they are ranked by profitability, then:

1. The forager should always include the most profitable prey in its diet.

2. Add additional prey types into the diet until the profitability of a new type is less than the net caloric intake rate of the diet without that prey type.

These rules allow you to test if an animal is foraging optimally. For example, consider the data in Table 14.1.

Table 14.1 Hypothetical diet for a fox.

Item	kcal	min	kcal/min	#/min
Meadow vole	160	3.2	50	0.2
Deer Mouse	25	0.6	42	3.0
Masked shrew	40	1.6	25	3.0

What is the optimal diet for this hypothetical fox? Meadow voles have the highest profitability (50 kcal/min) and should be included in the foxes diet according to rule #1. We can calculate the net intake rate for a diet consisting only of voles as:

The profitability of deer mice (42 kcal/min) is greater than the net intake rate of the vole only diet (19.5 kcal/min), so rule #2 says that deer mice should be

$$\frac{E}{T} = \frac{0.2(160)}{1 + 0.2(3.2)} = 19.5$$

included in the diet.

The profitability of masked shrews (25 kcal/min) is less than the intake rate from a diet of both voles and mice (i.e. 31.1), so shrews should not be added to the

$$\frac{E}{T} = \frac{0.2(160) + 3.0(25)}{1 + 0.2(3.2) + 3.0(0.6)} = 31.1$$

fox's diet. Voles and mice together are an optimal diet providing roughly 31.1 kcal/min of net energy. If the hypothetical fox was observed eating all three small mammals, then the hypothesis that the fox was foraging optimally would be falsified.

One additional point to make is that if the encounter rate increases, the net caloric rate for any diet increases, yet the profitability of each prey item remains the same. Can you explain why? The prediction is that if encounter rates increase, the optimally foraging animal should become more selective and eliminate one or more prey item from its diet. For the hypothetical fox example, encounter rates might increase as winter gives way to spring and summer, and the diet breadth of the fox should be narrower in the summer than in the winter. This prediction can also be tested with data collected in the field.

FORAGING IN PATCHES

In the wild, encounter rates usually vary across the habitat, and the habitat is said to be patchy. Thus, a given environment may have a range of patches from poor quality patches (where prey are absent or scarce) to high quality patches (where prey are diverse and abundant). Foragers move between patches looking for food. If patches vary in quality, then we can assign a profitability to each patch. This leads to a number of interesting questions:

· How many patches should a forager visit?

· How long should a forager spend searching within a patch?

· What is the optimum time to quite one patch for another?

A predator foraging in a patch will eventually deplete the food available in that patch and the encounter rate for prey will drop over time. Thus, if we plot the net caloric intake over time E *(t)*, we would expect a graph similar to Figure 14.3.

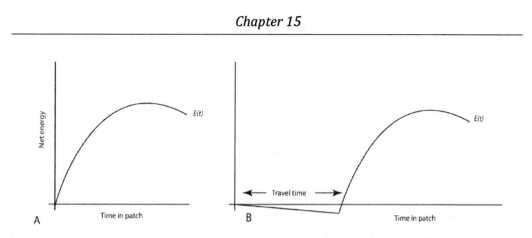

Figure 14.3   Graph of the net energy gain over time of foraging in a patch.

In Figure 14.3 part A, the forager gains a lot of energy after it enters the patch (e.g. the curve is increasing sharply), but as the forager spends more and more energy searching for fewer and fewer prey the curve begins to decline. The curve in part B is more realistic because here the travel time to get to the patch is added. Notice that travel has a cost associated with it and the first part of the graph dips below horizontal line indicating net energy lost while traveling.

The question is when should the predator quit foraging in this patch and move onto another patch? You might suggest that the predator should leave when the energy gain is maximized (e.g. at the highest point on the curve), but you would be incorrect. The reason is that the forager is presumably trying to maximize its average net caloric intake rate E/T, but in this case T includes travel time between patches and time spent within each patch. In other words, we are looking for the average rate of change. The maximum average net caloric intake rate is then the slope of the tangent line from the origin (Figure 14.4).

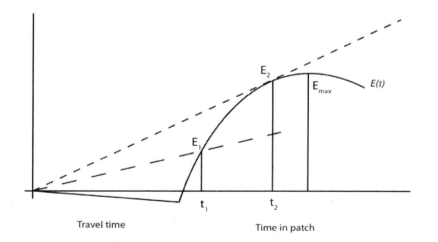

Figure 14.4   The average net caloric intake rate at three points in time. The E/T is maximized at the tangent to the curve.

If the forager quits the patch at time t_1 in Figure 14.4 , then the slope of the line from the origin to that point on the curve is less than the slope of the tangent line at time t_2. Since the slope of the line to any point on the curve is the average net caloric intake rate at that point in time, it should be obvious that *E/T* is maximized at the point on the curve where the tangent line passes through the origin (for details see Kolmes et al., 1997).

Note: This is called the marginal value theorem; it was borrowed from economics and first applied to behaviors by Charnov (1976).

The patch model can be tested by manipulating the travel time between patches. For example, if travel time between patches is lengthened, then it becomes costly to move between patches and the forager should remain longer in the patch before quiting and moving on to a new patch (Figure 14.5). This is illustrated by the lower slope of the tangent line in part B of Figure 14.5. We will test some of the predictions of optimal foraging theory in the exercises that follow.

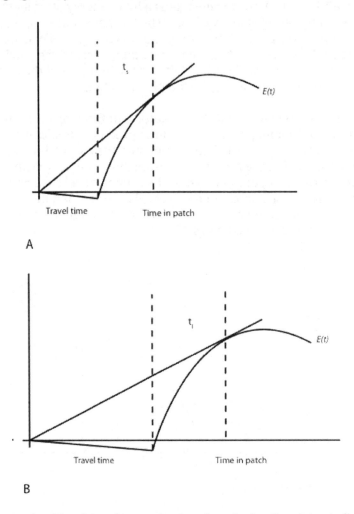

Figure 14.5 Two graphs of travel times between patches of equal value. Travel time is short in (A) and longer in (B)

EXERCISES

EXERCISE 1: PROFITABILITY AND PREY CHOICE

In this exercise you will present food items of varying energy content and handling times to foraging mammals. You will use squirrels as foragers (predators) and various nuts as prey.

PART 1:

1. Your instructor will provide you with two types of peanuts for your first trial. In this trail we will give the squirrels a choice between peanuts in the shell and peanuts with the shells removed. Because peanuts usually come two nuts to a shell, you will have to carefully cut the shelled nuts in half so that they contain a single nut. Why is this a critical step? Shelled peanuts have greater handling time because the squirrel must remove the shell to eat the energy rich nut inside. Therefore, you would predict that the squirrels should prefer the more profitable shell-less peanuts over the shelled peanuts.

2. Locate a place where squirrels are common and easy to observe.

3. Place a 2 foot by 2 foot white board near the squirrel's center of activity. (painting the tray white makes it much easier to observe when nuts have been removed).

4. On the white tray (board) place an equal number of shelled and shell-less peanuts.

5. Locate a place where you can sit and easily observe your subject squirrels but won't disturb their behavior.

6. Using binoculars, if needed, record how long each squirrel remains at the tray (patch), what nuts it eats or removes from the patch, and the order in which the two types of nuts are chosen.

7. Repeat the process several times to ensure you have enough replicates.

8. Summarize your results graphically, compare the removal rates of shelled and unshelled peanuts statistically, and write a short methods and results section in journal format (your instructor will provide details and resources).

PART 2:

1. Develop a new study that tests if squirrels forage optimally. Think about others ways to vary handling time or energy content of the food choices.

2. Create a hypothesis grounded in optimal foraging theory and a prediction about how squirrels should forage. Develop a study design that can test this prediction. Consult with your instructor prior to carrying out the study.

3. Conduct the study and carefully record your data.

4. Present your study design and results to the class.

EXERCISE 2: FORAGING IN PATCHES

In this experiment you will give squirrels a choice of two patches, a poor quality patch and a high quality patch. Patch quality will be set by the density of nuts. Each nut is assumed to be of equal caloric value.

1. Your instructor will provide each team with two feeding trays (each tray is 2 foot by 2 foot, with a 2 inch high perimeter; trays are filled with sand).

2. Count out 50 peanuts and randomly bury them in the sand in the first tray; this is the high quality patch.

3. Count our 10 peanuts and randomly bury them in tray 2; this represents the poor quality habitat.

4. Create a set of predictions for foraging time in patch, patch quitting time, and rate of foraging within each patch.

5. Set out the two patches within 5-6 feet of one another near the center of squirrel activity.

6. Use binoculars and a stop watch to record the squirrels foraging behavior and times.

EXERCISE 3: FORAGING WITH RISK

This experiment will test the foraging efficiency of grey squirrels under varying predation and competition risks. This experiment is similar to that of Newman and Caraco (1987) and it is a good idea to read that paper before you begin. The basic premise is that patches that are out in the open or far from cover are more risky places to forage because of the risk of predation. However, even patches near cover can be risky if there are many aggressive competitors (e.g. other squirrels) all trying to forage on the same food patches.

There are essentially four variables:

1) **Squirrel feeding rate**: In this experiment you use sunflower seeds because squirrels are more likely to eat them on the spot instead of caching them for later. You will record the number of seeds eaten per minute by each squirrel. (record the number of seeds and seconds even if the squirrel leaves the patch before a minute is up). If there is more than one squirrel foraging in the patch at one time, record each squirrel's feeding rate separately and the group size (see below). If group size changes during the minute, record the average group size for that minute. If a single squirrel stays for more than one minute, record it again for another minute (up to five minutes).

2) **Group size**: The number of squirrels foraging at the same time in the same patch (or nearby).

3) **Distance to cover**: The distance in meters to tree cover. The predation risk, energetic costs, and travel time all increases with distance to cover. You will place your food patches at either 5 or 15 meters from suitable cover.

4) **Patch size**: The number of seed-containing trays in each patch will be either 1 tray (small patch) or 4 trays (large patch).

Before you begin, think about the following questions and generate predictions that can be tested from the data you collect for the 4 variables just discussed.

· Would a squirrel prefer to forage in a large group or alone?

· Would group size likely be larger at greater distances to cover? Why?

· Given identical food items, would feeding rates increase or decrease with distance to cover ?

· How might patch size and distance to cover interact to affect group size and/ or feeding rate?

Your instructor will divide the class into groups and assign each group a patch type (small or large) and a distance (5 or 15 meters). One person in each group will be responsible for recording the data onto the data sheets, while the others will report the feeding rates and group sizes every minute. They should also note any aggressive behaviors (e.g. chasing).

1. Place your patch at the appropriate distance from tree cover and put an equal number of sunflower seeds in each tray.

2. Retreat to a place where you can observe but not disturb the squirrels and begin observations (be patient, you may have to wait a bit until the squirrels find the patch).

3. When you are finished, remove your patches and all seeds from the site. Summarize you raw data from you data sheet by calculating the means and standard deviations as shown in Table 14.2. Class data will have to be pooled from these summary sheets prior to statistical analysis.

Table 14.2 Summary data table for squirrel foraging.

Distance to cover =		Patch size =		
Group size	Mean Feeding Rate	Std.Dev. of Feeding Rate	Mean Group Size	Std. Dev. of Mean Group Size
1				
2				
3				
4				

For Feeding Rate data, calculate a mean and standard deviation for each group size for which there are observations. For Group Size observations over time, calculate an overall average group size across all times.

DATA ANALYSIS:

REGRESSION ANALYSIS:

Begin by looking for a relationship between the squirrel's feeding rate and 1) group size, 2) patch size, and 3) distance to cover. To do this you would create a scatterplot of the raw data (not the means) and plot a "best-fit" line to that data and provide the equation for that line with a constant slope. A linear regression determines the relationship between a dependent variable and an independent variable. Linear regression assumes 1) that there is a plausible reason to assume a cause and effect relationship between the variables, 2) that the data were randomly collected, and 3) that observations are independent (if you collected 5 sets of feeding rates all form the same squirrel, then these sets are not independent). The regression equation is the familiar equation for a line $y = mx + b$, where x and y are the data points, m is the slope of the line and b is the y-intercept. The regression equation has a constant slope m, meaning that the feeding rate changes at a constant rate as group size (or other variable) changes. The R^2 value is the percent of all the variance in one variable that can be accounted for by the linear regression model. If R^2 was 0.87 for a plot of squirrel feeding rates (y) versus group size (x), then 87% of the variance in feeding rate can be accounted for by group size. It also means that 13% of the variance remains unaccounted for. A p-value is also reported in most analyses of linear regressions. The p-value relates to the null hypothesis; in a linear regression the null hypothesis is that there is no relationship between the variables. If true, the regression line would be horizontal (slope = 0.0). If the p-value is less than 0.05 we can conclude that the null hypothesis is very unlikely to be true and that the observed relationship is not due to chance.

1. Use *Excel, OpenOffice*, or a statistics program (specified by your instructor) to plot the regressions of the variables you collected in you squirrel study.

2. Consider each regression carefully, what patterns do they suggest?

ANALYSIS OF VARIANCE (ANOVA):

Regressions are useful for determining the relationship between two variables (assuming the relationship is linear and not on a curve). However, if you want to know if the mean feeding rate for squirrels at large patches is different from those feeding at small patches, you need a different statistical test. You could use a student t-test, but an advantage of ANOVA over t-tests is that we can 1) test more than two means simultaneously, and 2) determine effects of more than one factor and any interactions between them. Suppose you want to test the effects of distance to cover and patch size on feeding rate and you also want to know if there is any interaction between patch size and distance to cover. In this case you need to use an ANOVA.

Your instructor will provide you with additional instructions on how to preform ANOVAs on the statistical platform at your institution. Use your ANOVA analysis to answer the following questions about your study:

1. Did you results support or falsify your initial hypothesis?

2. Which patch size gives the highest feeding rates?

3. Do feeding rates increase of decrease with distance from cover?

4. How do distance from cover and patch size interact?

5. How does group size change with distance from cover?

6. How does group size change with patch size?

7. Why might your data deviate from your predictions?

8. What selective factors may be at play in shaping the squirrels foraging behaviors?

BIBLIOGRAPHY

Charnov, E. L. (1976) Optimal foraging: attack strategy of a mantid. *American Naturalist*, 110:141-151.

Kolmes, S., Mitchell, K. and J. Ryan. (1997) Optimal foraging theory. *Journal of Undergraduate Mathematics and its Applications.* 18(1):43-85.

Krebs, J. R. (1978) Optimal foraging: decision rules for predators. In *Behavioural Ecology, An Evolutionary Approach*, (J. R. Krebs and N. B. Davies, eds.), Sinauer Associates, Sunderland, MA.

Newman J.A. & Caraco T. (1987) Foraging, predation hazard and patch use in grey squirrels. *Animal Behavior* ,35:1804-1813.

APPENDIX

Sample data table for squirrel foraging project

Distance to Cover: 5 or 15 m		Patch Size: Small or Large	
Time	Feeding Rate (#/min)	Group Size at time x	Aggressive Behaviors or Comments

15

Field Karyotyping

TIME REQUIRED

One 3-4 hour lab period for each of the exercises

LEVEL OF DIFFICULTY

Moderate

LEARNING OBJECTIVES

Understand the basics of mitosis and chromosome structure

Understand how karyotypes are made

Understand what kinds of information karyotype can yield

Learn how prepare a field karyotype

Learn how to collect and preserve tissues for DNA analysis

EQUIPMENT REQUIRED

Mitotic inhibitor (sterile vinblastine sulfate, USP), Eli Lilly and Company;

Eagles growth medium - GIBCO Laboratories;

Disposable 15-ml centrifuge tubes with caps, Corning Laboratory

0.075 M potassium chloride solution

Carnoy's fixative

Giemsa blood stain

Dehydration solutions (acetone and xylene)

Miscellaneous lab equipment, glassware and supplies:

Pasteur pipettes (6 inch),

Rubber latex bulbs,

1-cc insulin syringes,

Coplan staining jars,

Microscope slides (1.0 mm) and storage boxes,

Cover slips (24 by 50, No. 1 thickness),

Permount slide mounting medium,

Clinical centrifuge, or hand cranked field centrifuge

Gasoline power generator (for clinical centrifuge)

Centrifuge tube racks,

100-ml graduated cylinder,

Beakers,

BACKGROUND

Note: This lab assumes that you are familiar with the general structure and function of chromosomes and the basics of information transfer via DNA.

Mammalogists are often interested in questions that require an understanding or comparison of the chromosomes of one or more populations or species. Understanding the chromosomal complement of a mammal is useful for studying taxonomic relationships, chromosomal aberrations, cellular function, and to understand past evolutionary events. For example, the duckbilled platypus (*Ornythorynchus anatinus*, Monotremata) was recently discovered to have 5 pairs of sex chromosomes instead of a single pair (XX or XY) found in most mammals (Grutzner, et al. ,2004). Surprizingly, platypus sex chromosomes are strong homologous with bird sex chromosomes (not other mammals). This implies that the therian sex chromosomes evolved (probably from an autosomal pair) after monotremes 166 million years ago (Veyrunes et al., 2008).

The chromosomal complement, or **karyotype**, is typically uniform within a species. However, there may be considerable karyotypic variation among populations of a species. Thus, karyotypes often aid in our understanding of the relationships among species. For example, close examination of chromosomes reveals that rearrangement of chromosome segments into different combinations can explain much of the variation in species karyotypes. (for a review see Ferguson-Smith and Trifonov, 2007).

WHAT IS A KARYOTYPE?

A karyotype is either 1) the number and appearance of chromosomes in a eukaryotic cell, or 2) the complete set of chromosomes in an individual organism (or a species).

In essence, karyotypes describe the number of chromosomes, and what they look like (i.e. their length, centromere position, banding pattern, and any other physical differences). The chromosomes are arranged in pairs, by size (largest first) and, for similarly sized chromosomes, by position of centromeres, with the sex chromosomes positioned last (Figure 15.1). This standard format known as a karyogram (or idiogram). The short arm, called "p" (for petite), is always at the top, and the long arm, designated "q" (because it follows letter "p"), is always at the bottom.

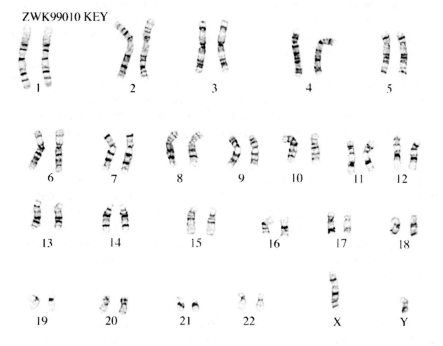

Figure 15.1 A karyogram of a male human showing the 23 homologous pairs of chromosomes (including an XY pair of sex chromosomes) stained with Giemsa. (Courtesy of Larry Phelps, University of Wisconsin Baraboo)

Karyograms illustrate the number of chromosomes in an individual or species' somatic cells; this somatic number is referred to as 2n (diploid). Gamete cells have a chromosome number of n (haploid). For example in humans 2n = 46, where n = 23).

HOW DO SCIENTISTS READ CHROMOSOMES?

Scientists typically read chromosomes in five ways. When scientists compare chromosomes from different species (or individuals from different populations), they look for differences in:

Size Differences: Chromosomes can vary in length by up to twenty times between genera within a family. Such size differences probably result from differences in number of DNA duplications. Alternatively, differences in relative size of chromosomes are caused by interchange of unequal lengths of segments of chromosomes.

Centromere Position: Centromeres are required for normal chromosome separation during mitosis or meiosis. Normally, each chromosome has a single centromere located near the middle of the chromosome. Such a position is termed **metacentric**. In contrast, some chromosomes have their centromere positioned near the very end of the chromosome (**acrocentric**) or off-center (**submetacentric**). Differences in the centromere position occur by **translocations** (rearrangement of parts between nonhomologous chromosomes).

Chromosome Number: Differences in basic number of chromosomes may occur due to repeated translocations which remove essential DNA from a chromosome by adding that DNA to other chromosomes. This eventually allows the loss of a chromosome without harm to the organism. For example, great apes have one more pair of chromosomes that humans, but the essential great ape genes were added (translocated) to other chromosomes in humans.

Satellite Position and Number: Some chromosomes have an additional constriction (besides the centromere) near the distal tip of the chromosome called a **satellite**. Satellites are small fragments attached to the main chromosome by a strand of chromatin. Differences in number and position of satellites can be useful in cytogenetics.

Heterochromatic Banding Patterns: Heterochromatin refers to regions of densely packed DNA which stains darker than euchromatin (lightly packed regions). In the karyotype in Figure 15.1, Giemsa staining reveals light and dark bands (G-banding). Giemsa stains phosphate groups of DNA. Each chromosome has a characteristic banding pattern, but both members of a chromosomes pair have identical banding patterns. Thus, differences in the relative staining of these regions can useful when comparing karyotypes.

Note: A common misconception is that bands represent single genes, but in fact the thinnest bands contain over a million base pairs and potentially hundreds of genes.

A full comparison of karyotypes may reveal variation between sexes, members of a population (polymorphisms), geographic regions (geographic races), or between individuals (abnormal individuals).

HOW ARE KARYOTYPES PRODUCED?

Conventionally, karyotypes were made by injecting the subject animal with mitotic inhibitors such as colchicine citrate or vinblastine sulfate, which arrests cell division at metaphase when chromosomes are condensed. After a 24 hour period the animal is sacrificed and the bone marrow removed and placed on a microscope slide. This technique, which uses live animals, has modified for use with postmortem samples (dead animals). The postmortem technique described below is more appropriate for field work where housing live animals overnight is difficult or impossible.

Note: Be sure to follow the procedures spelled out in your approved IACUC protocol and the Guidelines of the American Society of Mammalogists for the use of wild mammals in research, which can be found at:

http://www.mammalsociety.org/whats-new/new-animal-care-and-use-guidelines

FIELD KARYOTYPE PROCEDURES

EXERCISE 1: FIELD KARYOTYPING

Following the capture of an animal (see Chapter 5), the animal is euthanized (alternatively, animals euthanized as part of another study can also be used) and the following information is recorded.

1) sex: male/female

2) age: adult/immature

3) for females, record whether lactating or pregnant

4) weight to nearest 0.1 g

5) head + body length to nearest 0.1 mm

6) tail length to nearest 0.1 mm

7) length of right hind foot to nearest 0.1 mm.

PROTOCOL

1. Immediately after euthanasia, expose and remove the femur. Cut the head of the femur off near its proximal tip to expose the bone marrow cavity.

2. Using a narrow gauge needle attached to a syringe filled with Eagles medium, insert the needle into the marrow cavity and flush the marrow contents into a 5 ml disposable tube. Eagles medium (Minimum Essential Medium [Eagle], with Hanks' salts and HEPES Buffer) from GIBCO Laboratories.

3. Gently break up any marrow clumps with a 23-gauge needle (or by drawing the marrow solution back and forth into the syringe).

4. To each 5 ml of media-marrow suspension, add 0.1 ml of a 0.001% colchicine solution (mitotic inhibitor) or 0.01% Velban (sterile vinblastine sulfate), shake, cap the tube, and incubated at 37°C for 90 min. Under field conditions incubate samples close to one's own body.

5. Centrifuge the tube for 3 min at approximately 3,000 rpm using a manual (hand cranked) or battery-operated centrifuge until a pellet forms in the bottom of the tube.

6. Carefully remove and discard the supernatant (leaving cell pellet) and add approximately 3 ml of hypotonic solution (0.075 M potassium chloride)

7. Resuspend cells in hypotonic solution by aspirating cell pellet with a pipette.

8. Recap the tube and incubate at 37°C (near body) for 15 min.

9. Centrifuge the tube for 3 min at approximately 3,000 rpm.

10. Carefully discard supernate (leaving cell pellet), add approximately 3 ml of fixative solution, and resuspend cells in fixative. Carnoy's fixative is one part glacial acetic acid and three parts absolute methanol (made fresh before use). This is Wash 1.

11. Centrifuge for 3 min at approximately 3,000 rpm.

12. Discard the supernatant fixative, add approximately 2 ml of new fixative, and resuspend cells.

13. Aspirate entire cell suspension in a pipette and filter through cheesecloth into the original centrifuge tube. If necessary add more fixative to bring volume of suspension to approximately 3 ml. This is Wash 2.

14. Centrifuge for 3 min at approximately 3,000 rpm.

15. Discard supernatant fixative, add approximately 1 ml of fresh fixative, and resuspend cells. The solution should be visibly cloudy.

16. Place a clean microscope slide on the benchtop. Allow 3-5 drops of the cell suspension to drop from a distance of 1-2 feet above a clean microscope slide. This allows the chromosomes to spread out a bit and make them easier to see. Allow the preparation to air dry.

17. Insert the microscope slide into a coplin jar containing Giemsa stain for 15 min. Giemsa stain is one part stock Giemsa with eight parts hot tap (or distilled) water (made fresh before use).

18. Remove slide from Giemsa and pass it through a series of dehydration baths. Dehydrate quickly, 2 dips, 1 second each, in two baths of acetone, one bath of acetone and xylol (1:1), and two baths of xylol.

19. Coverslip immediately with 3-4 drops of Permount.

EXERCISE 2: G-BANDING CHROMOSOMES WITH TRYPSIN

This protocol may provide superior G-banding results as it uses an enzymatic treatment of the slides. Investigators should adjust the times to suit the slide quality desired. This exercise begin after the slide has air dried in step 16 above.

1. Age the air-dried slide overnight in a drying oven at 55^0 to 60^0C. Remove the slides and bring to room temperature just prior to banding.

2. Grasp the slide with forceps and immerse it in a Coplin jar containing the 0.025% trypsin working solution for 8 to 10 seconds, moving the slide back and forth gently.

Trypsin working solution 0.025% is: 0.5 g trypsin (Difco, 0.025% w/v final) in 200.0 ml Earle's balanced salt solution. Store up to 1 day at room temperature.

---

3. Briefly rinse slide 1% FBS to inactivate the trypsin.

4. Pre-rinse slide by dipping in Gurr's buffer, using the same agitation technique as in step 2

Gurr's buffer solution, pH 6.8: dissolve 1 Gurr's buffer tablet in 1 liter sterile H_2O.

---

5. Place slide in a Coplin jar containing Giemsa staining solution for 8 to 10 min. Giemsa stain working solution

---

Giemsa staining solution: 1 ml Giemsa stain (Azure Blend, Harleco 620G/75, from EM Science; 2% v/v final) added to 50 ml Gurr's buffer solution, pH 6.8: Store up to 2 months at room temperature

---

6. Rinse slide in sterile distilled water until the stain no longer discolors the water, using the same agitation technique as in step 2.

7. Allow slide to air dry. Examine by light microscopy using a phase-contrast microscope to determine the quality of banding. Adjust trypsin exposure or duration of staining as required.

8. Once the optimal banding quality has been achieved, coverslip, and analyze the slides.

EXERCISE 3: ANALYZING THE KARYOTYPE MANUALLY

Photograph (digitally) appropriate spreads and produce 8 X 10 high contrast photographs of your chromosome spreads and print these images out.

Cut each chromosome from the photograph and arrange the chromosomes according to size and position of the centromere (this can also be done digitally using Adobe Photoshop). It's best to do this one at a time rather than cutting them all out at once and then trying to match them (you're more likely to lose one).

This may be difficult if the chromosome spreads on the slides are of poor quality. In this case, use the sample human chromosome spread in the Appendix A and B (provided courtesy of Larry Phelps). Carefully cut out each chromosome and compare each chromosome to its potential homolog in partially completed human karyotype in Appendix B.

Tape or glue each chromosome next to its pair on a form supplied for this purpose.

The analysis involves comparing chromosomes for their length, the placement of centromeres (areas where the two chromatids are joined), and the location and sizes of G-bands.

Determine the diploid number of chromosomes, sex, and number of metacentric, submetacentric, and acrocentric chromosomes.

If possible compare the karyotypes from several mammalian species and note any differences in the karyotypes.

Analyzing karyotypes by hand is time consuming. It is now easier to do it using image analysis software on a computer.

EXERCISE 4: MEASURING CHROMOSOMES

WITH *IMAGEJ* SOFTWARE

You need to create digital images of the karyotypes from your preparations in exercise 1 and 2. Your instructor will show you how to use the digital camera attached to your microscope. (alternatively, digital images of human karyotypes can be found on the web at:

http://worms.zoology.wisc.edu/zooweb/Phelps/karyotype.html

Before you begin taking photos of your chromosome preparations, you will need to take a photograph of a micrometer slide so that we can set the scale of our images. A stage micrometer has a finely divided scale etched on the surface (Figure 15.2). The scale is a precise true length and is used for calibration.

Place the micrometer slide on the microscope stage. Under the lowest magnification, adjust the focus so that the etched lines are in focus. Take a digital photo and save it with an appropriate title (include the magnification in the title –i.e. micrometer 4X). Repeat this procedure for the other objectives you plan to use (be very careful when switching objectives that you do not scratch the lens across the micrometer slide).

Remove the micrometer slide and replace it with the chromosome preparation slides you wish to measure. Locate regions of interest and photograph, and save them. Record each image with a unique and informative label and save them to a folder.

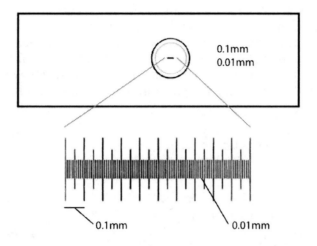

Figure 15.2 A drawing of a stage micrometer (top) showing the scale ruler at the center and the dimensions of the large and small lines etched on the slide (right), and a close up of the scale (bottom).

Note: each microscope is slightly different and it is good practice to take micrometer photos at each power for each microscope.

You will apply image analysis techniques to your digital images, using the public domain software *ImageJ*, written and maintained by Wayne Rasband at the National Institute of Mental Health, Bethesda, Maryland, USA (ImageJ is the successor of NIH Image). *ImageJ* is written in Java, which means that it can be run on any system for which a java runtime environment (JRE) exists (i.e. Windows, Mac, Linux).

If you haven't already done so, go to the *ImageJ* website and download the appropriate version for your operating system. http://imagej.nih.gov/ij/.

Note: Java applications will only use the memory allocated to them. Under *Edit Options>Memory & Threads...* you can configure the memory available to ImageJ. The maximum memory should be set to 3/4 of the available memory on your machine.

In the first part of this exercise you will use the photo of the micrometer slide to create a scale bar for each subsequent image. Suppose you are using the photo of a micrometer slide taken at 40X.

1) Begin by opening ImageJ and then opening the image file for the 40X micrometer slide taken earlier (i.e. *File>Open* and then browse to the correct image).

2) Select the straight line tool from the *ImageJ* toolbar (5th from the left, Figure 15.3) and place the cursor (which is a crosshair) carefully at the start of one major line, click and drag the crosshair to the next major line (or over several major lines). Recall that the distance between major lines is recorded on the surface of the micrometer slide (the larger of the two numbers). For example, if you micrometer slide is marked 0.1mm and 0.01mm, then the major lines are each 100 microns apart and the smaller lines are each 10 microns apart. Record the length in pixels for the distance you measured (it is given just under the imageJ toolbar in pixels. For example, if there are 1865 pixels in 0.1 mm (100 microns), you would record 1865/100um (or 18.6 pixels per micron) in your notes. Repeat this for each magnification.

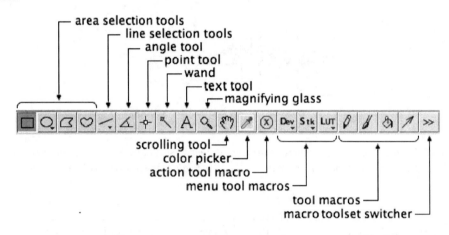

Figure 15.3 A close up of the ImageJ toobar with tools listed.

3) In *ImageJ* go to *Anaylze>Set Scale*. Enter the distance in pixels from your measurement (i.e. 1865 in this example) and the know distance (i.e. 0.1 in this example) and the unit of lenth (i.e. mm). Check the *Global* box to apply this scale to all images in the current ImageJ session.

Note, if you restart the program, you have to re-enter the scale each time, so have all your ratios written down.

---

4) Click *OK*, and test the settings by re-measuring the major lines and making sure that they are 100 um apart.

Note: use the keyboard shortcut ctrl+m

---

5) Open your first chromosome image (*File>Open*). Add a scale bar to the image by going to *Analyze>Tools>Scale Bar*. Specify the width, thickness, font size, color and location of the scale bar and click *OK*.

Now you are ready to measure some chromosomes. You can do this in two ways. First you can use the tools already present in *ImageJ* to measure each chromosome. Set the scale as described above using *Analyze/Set Scale*.

6) Use the magnifying tool to enlarge the region of interest. Click on the line segment tool to select it. (Line segment tool is found by right clicking the straight line tool – 5th from left in tool bar).

7) Move the mouse cursor over the chromosome and click and drag from one tip to the middle and click there once, then continue moving the cross hair to the other tip and double click. You should now have line with two segments that runs the length of the chromosome (Figure 15.4).

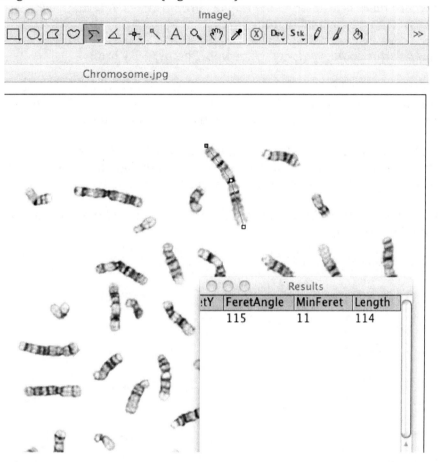

Figure 15.4 A view of the ImageJ windows showing the icon for the line segment tool in the toolbar, a chromosome with a line segment over it, and a results window showing the length of the chromosome as 114 microns.

8) Use the *Analyze/Measure* command to measure the length of the chromosome line. The measurement show up in the Results window under length (see Figure 15.4). Measurements can be transferred to a spreadsheet by right-clicking in the Results window, selecting *Copy All*, switching to the spreadsheet program, and then pasting.

The second way to measure chromosome arm lengths is to use a plugin written specifically for this task called **LEVAN**, developed by Yugo Sakamoto and Adilson Zacaro and available at:

http://rsb.info.nih.gov/ij/plugins/levan/levan.html

Visit this site, download the *Levan.jar* file, and move it to the plugins folder in ImageJ, then restart ImageJ and the Levan command will appear in the drop down list under *Plugins* in the menu bar.

9) Close any open images. Go to *Plugins>Levan* to open the Levan plugin in a new window.

10) Click the *Open* button and browse to the image of your chromosomes. Use an image where there is little or no overlap of chromosomes. Alternatively, use one of the processed images available from:

http://worms.zoology.wisc.edu/zooweb/Phelps/46xy.html

11) *Levan* opens the karyotype image in a new window. Choose the line selection tool on the tool bar and with the mouse trace each arm of each chromosome. As you do this, each arm will be added to the list in the *Levan* window.

12) Click on *Classify* in the Levan window to classify each chromosome as metacentric, submetacentric, etc. Levan also gives you the length of the p and q arms in pixels. You will need to convert this to microns manually.

13) Click on the *Draw* button in the *Levan* window to add the numbered arms to the photo (Figure 15.5).

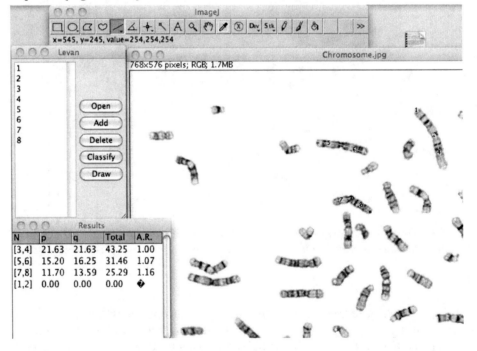

Figure 15.5 The Levan windows in ImageJ showing several marked chromosome arms and the results table.

14) To save the data to another location, go to *File>Save As* in the *ImageJ* menu.

BIBLIOGRAPHY

Bulatova, N.Sh., Searle, J.B., Nadjafova, R.S., Pavlova, S.V., and N.V. Bystrakova. (2009) Field protocols for the genomic era. *Comparative Cytogenetics*, 3:57-62.

Christidis, L. (1985) Rapid procedure for obtaining chromosome preparations from birds. *Auk*, 102:892-893.

Ferguson-Smith, M.A., and V. Trifonov. (2007) Mammalian karyotype evolution. *Nature*, 8:950-962.

Grutzner, F., Rens, W., Tsend-Ayush, E., El-Mogharbel, N., O'Brien, P.C.M., Jones, R.C., Ferguson-Smith, M.A., and J.A. Marshall Graves. (2004) In the platypus a meiotic chain of ten sex chromosomes share genes wth the bird Z and mammal X chromosomes. *Nature*, 432, 913-917

Hafner, J.C., and D.R. Sandquist (1989) Postmortem Field Preparation of Bird and Mammal Chromosomes: An Evaluation Involving the Pocket Gopher, *Thomomys bottae*. *The Southwestern Naturalist*, 34:330-337.

Veyrunes F., Waters, P.D., Miethke, P., Rens, W., McMillan, D., Alsop, A.E., Grützner, F., Deakin, J.E., Whittington, C.M., Schatzkamer, K., Kremitzki, C.L., Graves, T., Ferguson-Smith, M.A., Warren, W., and J.A. Marshall Graves. (2008) Bird-like sex chromosomes of platypus imply recent origin of mammal sex chromosomes. *Genome Research*. 18(6):965-973.

APPENDIX A

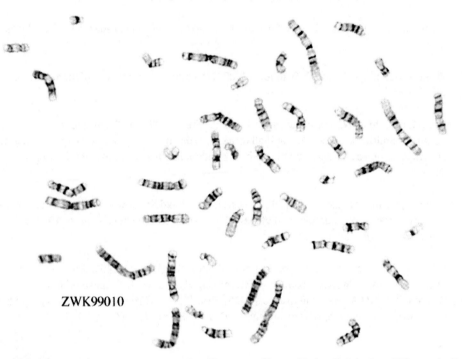

ZWK99010

Figure 15.3 A human chromosome preparation (Courtesy of Larry Phelps, University of Wisconsin, Baraboo)

APPENDIX B

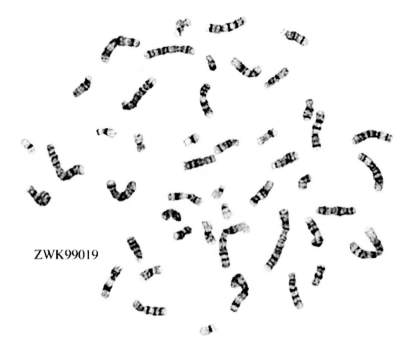

ZWK99019

Figure 15.4 A human chromosome preparation (Courtesy of Larry Phelps, University of Wisconsin, Baraboo)

16

Non-invasive Hair Sampling

TIME REQUIRED

One 3-4 hour lab period for each of the exercises

LEVEL OF DIFFICULTY

Basic to Moderate

LEARNING OBJECTIVES

Understand the basics of hair morphology

Understand how hair samples can be collected in the field

Understand how reference collections are made

Learn how to analyze hairs using ImageJ software

Learn how to preserve hair samples for DNA analysis

EQUIPMENT REQUIRED

PVC tubing of various diameters

Double sided sticky tape or rodent glue boards

Raw meat (bait) enough for every trap station

Thumbtacks (handful)

Hair sample envelopes (~20)

Permanent markers

Cordless drill/screwdriver and exterior wood screws

GPS with location of the station

Wooden boards (2 per hair station)

ImageJ software (free OpenSource)

Microscope slides and coverslips

DPX Mounting medium or Permount

Reagents and stains (described below)

Nail polish (clear)

Microscope

Coplan jars

QIAamp DNA kit

Microcentrifuge

Adjustable micropipettes and disposable tips

Vortex shaker.

BACKGROUND

Mammalogists are often interested in surveying the distribution and abundance of species within a region. While, it may be possible to live trap and identify individuals, it is often time consuming and expensive if the area is large or the species are rare or elusive (e.g. many carnivores).

Non-invasive techniques have been developed for carnivores (Long et al., 2008) and small mammals (Pocock and Jennings, 2006) as alternatives to traditional trapping. These non-invasive techniques include remote camera systems (see chapter 9), the use of track plates, and hair traps. Each method has its pros and cons and combinations of methods are needed to meet some study objectives.

Hair sampling has gain popularity as a non-invasive method for sampling species distribution or occupancy (see Kendall and McKelvey, 2008). In addition, hair sampling combined with molecular techniques has used to measure population density, monitor mammal populations, and increasingly to develop management strategies for wide ranging or elusive species.

Hair traps are now routinely used in large-scale field surveys to where the goal is to detect small to medium-sized terrestrial mammals. Hair traps come in a variety of shapes, sizes, and configurations. They are described as hair snares, hair tubes, sticky traps and other names depending on the specific design. For example, hair tubes require an animal to enter a tube and pass by sticky tape, which collects a hair sample, before reaching bait at the far end of the tube. Hair snares, by contrast, work by collecting hairs on a wire or stiff brush as the animal rubs against it.

Regardless of the method used to collect the hair samples, the hairs left behind can be analyzed by microscopic examination or DNA methods to reveal the species that deposited the hairs.

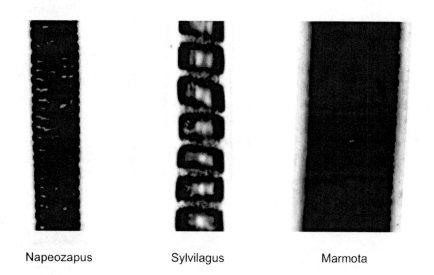

| Napeozapus | Sylvilagus | Marmota |

Figure 16.1 Microscopic images of mammalian guard hairs from a jumping mouse (Napeozapus), rabbit (Sylvilagus), and a woodchuck (Marmota). All images taken at the same scale.

HAIR MORPHOLOGY

Hair is a diagnostic character for the Mammalia. It is associated with the integument and grows out of a structure called a follicle (see Vaughan et al, 2010). The root of the hair lies within the follicle, and the portion above the skin surface is the hair shaft. The bulb forms the base of the root. Structurally, hair is composed of interconnected keratin proteins in the form of long fibrils. Keratin makes hair highly resistant to wear and chemicals. This is because keratin proteins have many cross-linages and disulfide bonds (S-S) link adjacent protein chains.

In essence, a hair is a long thin tube (Figure 16.2). The hair shaft consists of a translucent outer layer called the **cuticle**, consisting of overlapping cells that resemble the scales on a snake. These cuticle cells keratinized (hardened) and contain no nuclei and no pigments. The scales overlap and have their free end pointing distally. Cuticle patterns can be helpful in identifying species, but these patterns may vary even within an individual depending on the location of the hair (e.g. a guard hair may have a different cuticle pattern from underfur).

Just inside the cuticle is the **cortex.** It is composed of long, keratinized filaments running parallel to the hair shaft. Trapped within these filaments are air spaces called **fusi**. Hair color is primarily due to the density of pigment granules (melanin) in the cortex.

At the very core of the hair shaft is the **medulla**. The medulla varies from individual to individual and between hairs of a given individual. The medulla may be continuous throughout the entire hair shaft, fragmented, or even absent altogether.

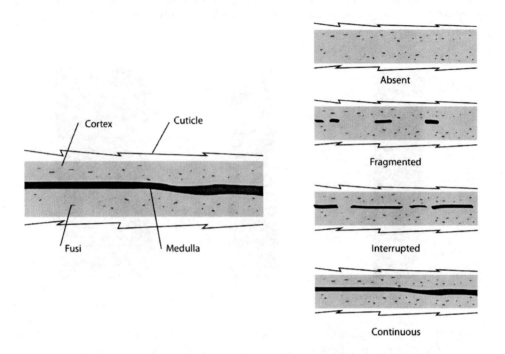

Figure 16.2 Diagrams showing the interior of a hair shaft cut longitudinally.

Many mammalian hair shafts show variation in diameter along their length. For example, the base may be smaller in diameter than the mid-section and the tip may taper to a blunt point.

A number of descriptive terms have been developed to aid in the analysis of hairs by forensic scientists and wildlife biologists. For example, Figure 16.2 illustrates that the medulla may be fragmented, interrupted, continuous, or absent. Likewise, there are terms to describe the cuticular scale patterns (Figure 16.3).

EXERCISES

EXERCISE 1:FIELD METHODS FOR COLLECTING HAIRS

There are a large number of devices available for collecting hair in the field. Among the most popular are hair tubes (of one kind or another) and hair snares. Hair snares use barbed wire or metal brushes (e.g. gun-cleaning brushes) to snag the hairs as an animal rubs against the snare. Hare snares are usually used for larger mammals such as ungulates and carnivores. For small mammals, the most commonly used device is a hair tube fitted with sticky tape. Hair tubes are baited and set out in areas were the target species are likely to occur. Hair tubes are made in a variety of shapes and sizes.

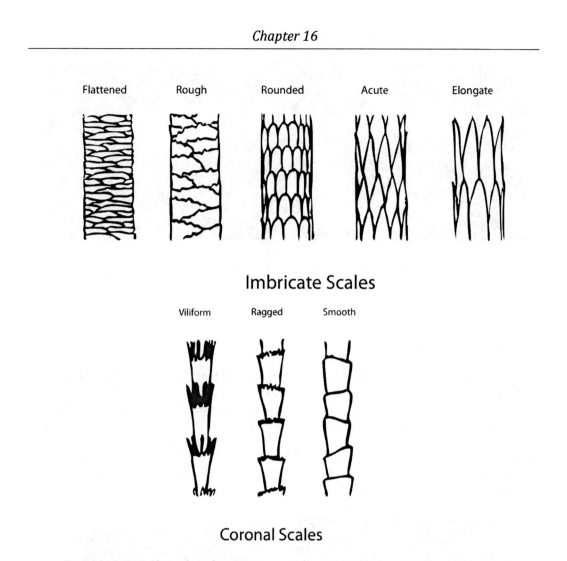

Imbricate Scales

Coronal Scales

Figure 16.3 Typical cuticle scale patterns seen in mammalian hair. Imbricate scales overlap in rows, while coronal scales encircle the shaft.

In this lab, you will make your own hair tubes and set up a collecting transect for surveying small mammals.

SHREWS AND RODENTS

Hair tubes for small mammals need to be made in various sizes. Cut lengths of PVC plastic pipe into 12 in (30 cm) lengths. Use several diameter pipes to cover the range of small mammal body sizes you expect to encounter. For example, cut 1 length each of 4 in, 3 in, 2 in, and 1.5 in diameter PVC. Line the inside wall on one side with one or more strips of double sided carpet tape, leaving one side of the pipe uncovered to act as the floor. (Figure 16.4). It is useful to cut the PVC tube in half lengthwise; the two halves are then duct-taped back together. This allows easy access to the interior of the tubes when removing hair samples and replacing sticky tape in the field. Cut wooden plugs to fit into the PVC at one end. These plugs are fitted on one side with wire mesh tea infuser balls that hold the bait. Traps are baited with a combination of rolled oats, honey and peanut butter. The ends of the PVC are plugged at the site with the bait facing the inside of the tube. Baited hair tubes are placed next to each other on the ground near potential runways.

Figure 16.4 Hair tubes made of PVC pipe. Left, hair trap strapped to a tree branch with Velcro straps. Right, set of 3 hair tubes set along a downed log.

SMALL CARNIVORES

The following hair sampling device is modified from those developed by Mowat and Paetkau (2002) American marten (*Martes americana*). The traps consist of two pine boards joined together lengthwise to form a roof. Each board is 1in x 6 in x 24 in (2- x 14- x 60-cm); the two are glued and screwed together along their long axis to from a 90⁰ angled roof. The two pieces of pine screwed together lengthwise along the edges to form a roof (Figure 16.5). Short strips of double-sided carpet tape or Catchmaster® rodent glue boards are attached to the underside of the roof. Glue strips are attached near both ends of the trap; the center section is used for bait or scent lures.

Figure 16.5 Wooden hair trap used for small carnivores. Sticky tape is affixed to the underside of the triangle. Left, hair trap set vertically on a tree trunk. Right, hair trap set horizontally on the ground.

The trap is either screwed to a tree trunk at chest height or placed on the ground in suitable cover. The traps form a triangular tent through which the animal can enter. Traps are baited with raw meat or other suitable bait for the target species (e.g. peanut butter and oats for squirrels)

DEPLOYING AND CHECKING HAIR TUBES

Twenty sets of PVC hair tubes (a set includes all diameters) are placed in suitable habitat along a transect. Sets are placed at roughly 20 meter intervals, numbered, and flagged. The number of tubes used for each site in any survey should be standardised. Traps can be attached to vertical tree trunks using screws or on horizontal branches using Velcro tape or hose clamps around the trap and branch (Fig. 16.4).

Checking traps: traps are carefully checked by 1) unscrewing the wooden carnivore traps from the tree and observing the glue boards for hair. If a board has hair, remove the entire glue strip (or tape) and place it in a sample bag or vial labeled with the station, location, your initials, date, time, and trap position (tree trunk, tree branch, ground, etc.) Hair tubes are checked by removing the wooden plugs and duct tape to separate the two halves. Tape is removed and replaced (if needed). All trap stations are checked for bait and the bait replaced if necessary. Traps are reassembled and put back in their previous locations.

EXERCISE 2: CREATING A HAIR REFERENCE COLLECTION

Creating microscope slides of hair samples is relatively easy. The basic set up (Figure 16.6) includes several coplan jars (i.e. jars that hold slides while they are staining or rinsing), microscope slides, coverslips, and mounting material to attaching the coverslips (e.g. Permount® or DPX mountant). .

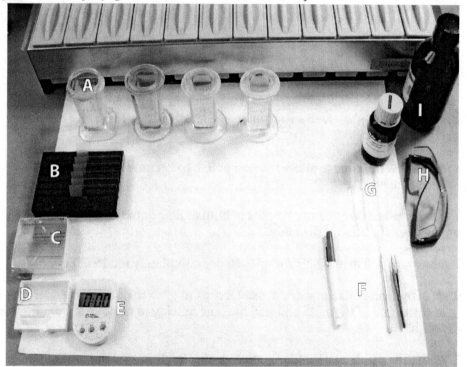

Figure 16.6 Typical staining set up. A- coplan jars, B-slide rack, C- microscope slides, D- 40mm cover slips, E-digital timer, F-pen and forceps, G-disposable pipettes, H- safety glasses, I-coverslip mount (e.g. DPX Mountant or Permount®)

Two types of preparations are necessary. The first is a cast of the outer cuticle that will be used to identify the cuticle scale type. Casts are made by placing the guard hair in a semi-dry bed of clear nail polish and then removing it when the nail polish has dried. What remains in the cast of the cuticle.

SCALE CAST PREPARATIONS

1. Place a drop of clear nail polish at one end of a clean microscope slide.

2. Place a coverslip next to the drop of fingernail polish, allowing the drop to run along the edge of the coverslip, and tilt the coverslip to a 45⁰ angle and lightly drag the coverslip across the slide to spread the polish evenly across the surface (Figure 16.7).

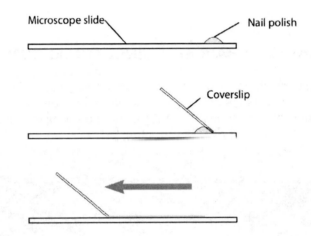

Microscope slide — Nail polish

Coverslip

Figure 16.7 Diagram of how to create an even coating of nail polish for preparing a hair cast.

3. Wait a few minutes to allow the nail polish to become slightly tacky. The timing takes some practice.

4. Quickly place three or four pieces of human hair onto the polish. Leave each hair hanging over the edge of the slide.

5. Allow nail polish and hair samples to dry completely for 15-20 minutes..

6. After the nail polish has dried, use forceps to grab the part of the hair that is hanging over the end of the slide and pull the hair quickly to remove it from the slide and nail polish.

7. View the sample casts under a microscope. Begin under the lowest power and focus on a hair cast, then switch to 10X or 40X lens to view the details.

8. Be sure to return the lens to the smallest lens before removing your slide from the microscope stage.

9. Illustrate the scale patterns you see in your notebook and label them with the term from Figure 16.3 that best describes the hair cast.

WET MOUNT PREPARATIONS

1. Label a clean microscope slide with the species name of the hair sample you will be working with.

2. Using a q-tip swab, wet a small surface of the slide with the DPX mounting medium (or Permount). Place the hair sample in this wet area to attach the hair to the slide. Hairs should be mounted horizontally and, if small, multiple hairs can be used on the same slide (avoid overlap).

Note: long hairs may require longer coverslips

---

3. Gently add several drops of mounting medium to the microscope slide (avoid moving the hair).

4. Touch one end of the cover slip onto the slide at the edge of the pool of mounting medium. Tilt the coverslip to a 45^0 angle (towards the hair sample) and slowly lower the coverslip down onto the mounting medium (i.e. go from 45^0 to 30^0, then to 15^0 and so on). Gently apply pressure to the other end of the coverslip (where it first touched the medium). This forces air bubbles out ahead of the cover-slip and make a much nicer preparation.

5. If the mounting medium does not completely fill the cover slip, add more along the coverslip edge.

6. Allow the slide to dry completely. If using DPX mountant this will be only 15 minutes or so (especially if you place the slide in a drying oven (use a low temperature setting).

7. Once the mountant is dry, scrape any excess mountant off the slide/cover-slip using a single edge razor blade (this is more difficult if you are using Permount).

8. Observe the slides under a light microscope at approximately 40x.

9. Observe and record the medullary and cortex characteristics.

10. Estimate the medullary index = width of medulla/width of hair.

11. Take digital photographs of each hair sample for use in the analysis described in exercise 3.

EXERCISE 3: QUANTIFYING HAIR STRUCTURE

USING *IMAGEJ* SOFTWARE

ImageJ software is introduced in Chapter 15, exercise 3. The basic procedures are similar for hair analysis. A brief description of the analysis is given below.

Before you begin the analysis you will need digital photos of your hair samples and digital photograph of a micrometer slide taken at the same magnification as your hair photos (see Figure 15.2).

You will apply image analysis techniques to your digital images, using the public domain software ImageJ, written and maintained by Wayne Rasband at the National Institute of Mental Health, Bethesda, Maryland, USA (ImageJ is the successor of NIH Image). ImageJ is written in Java, which means that it can be run on any system for which a java runtime environment (JRE) exists (i.e. Windows, Mac, Linux).

If you haven't already done so, go to the ImageJ website and download the appropriate version for your operating system. http://imagej.nih.gov/ij/.

Note: Java applications will only use the memory allocated to them. Under Edit Options>Memory & Threads... you can configure the memory available to ImageJ. The maximum memory should be set to 3/4 of the available memory on your machine.

In the first part of this exercise you will use the photo of the micrometer slide to create a scale bar for each subsequent image. This procedure is described in Chapter 15 exercise 3. With the scale set properly, you are ready to take measurements of your hair samples.

1) Open your first hair image (*File>Open* and browse to the correct file). Add a scale bar to the image by going to *Analyze>Tools>Scale Bar*. Specify the width, thickness, font size, color and location of the scale bar and click *OK*.

2) Use the magnifying tool to enlarge the region of interest if necessary. Click on the line segment tool to select it. (Line segment tool is found by right clicking the straight line tool – 5th from left in tool bar).

3) For hair diameter, move the mouse cursor over the hair and click and drag from one outside edge across the hair at 90^0 to the opposite outside edge and double click to release the line segment (Figure 16.8).

4) Create a data table of measurements for total diameter, medulla thickness, cortex thickness, etc.

5) Measurements can be transferred to a spreadsheet by right-clicking in the Results window, selecting *Copy All*, switching to the spreadsheet program, and then pasting.

Additional information about microscopic examination of hair can be found in Deedrick and Koch (2004).

EXERCISE 4: EXTRACTING DNA FROM HAIR SAMPLES

Forensic scientists routinely extract DNA from hair samples collected a crime scenes. Wildlife biologists have adapted these techniques to answer questions about species identify, sex, and paternity from hair or feces collected in the field. While, there are a number of techniques for extracting DNA, this exercise will introduce the technique using a relatively simple kit from QIAGEN, Inc. Santa Cruz California. (Promega also make a similar kit called DNA IQ™).

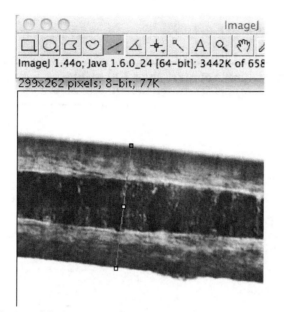

Figure 16.8 A view of the ImageJ windows showing the icon for the line segment tool in the tool-bar, a hair sample with a line segment over it, and a results window showing the diameter of the hair in microns.

Remove hair from glue plates or tape by washing in xylene until glue softens and hairs can be removed. Follow the procedures described in the kit. The basic outline is given below.

1. Cut 15 hairs into small pieces, place in a 1.5 ml microcentrifuge tube, and add 200 µl Buffer X1. Incubate at 55°C for at least 1 h until the sample is dissolved. Invert the tube occasionally to disperse the sample, or place on a rocking platform.

2. Add 200 µl Buffer AL and 200 µl ethanol to the sample and mix by pulse-vortexing for 15 s. After mixing, briefly centrifuge the 1.5 ml microcentrifuge tube to remove drops from inside the lid.

3. Carefully apply the mixture from step 2 (including the precipitate) to the QIAamp Mini spin column (in a 2 ml collection tube) without wetting the rim. Close the cap, and centrifuge at 6000 x g (8000 rpm) for 1 min. Place the QIAamp Mini spin column in a clean 2 ml collection tube (provided), and discard the tube containing the filtrate.*

4. Carefully open the QIAamp Mini spin column and add 500 µl Buffer AW1 without wetting the rim. Close the cap, and centrifuge at 6000 x g (8000 rpm) for 1 min. Place the QIAamp Mini spin column in a clean 2 ml collection tube (provided), and discard the collection tube containing the filtrate.*

5. Carefully open the QIAamp Mini spin column and add 500 µl Buffer AW2 without wetting the rim. Close the cap and centrifuge at full speed (20,000 x g; 14,000 rpm) for 3 min.

6. Place the QIAamp Mini spin column in a clean 1.5 ml microcentrifuge tube (not provided), and discard the collection tube containing the filtrate. Carefully open the QIAamp Mini spin column and add 200 µl Buffer AE or distilled water. Incubate at room temperature for 1 min, and then centrifuge at 6000 x g (8000 rpm) for 1 min.

7. Repeat step 6 (see also Fig. 16.9).

Note: Do not elute volumes of more than 200 µl into a 1.5 ml microcentrifuge tube.

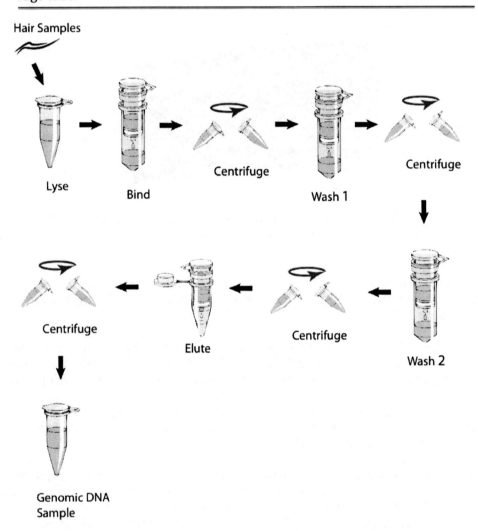

Figure 16.9 Sample protocol for extraction of DNA from hair samples using QIAamp™ DNA Mini Kits.

Note: For long-term storage, elute DNA in Buffer AE and store at –20°C.

LIBRARY, UNIVERSITY OF CHESTER

BIBLIOGRAPHY

Deedrick, D. and S. Koch. (2004) Microscopy of Hair Part II: A Practical Guide and Manual for Animal Hairs, *Forensic Science Communications*, Vol 6, (Available: http://www.fbi.gov/hq/lab/fsc/backissu/july2004/research/2004_03_research02.htm.)

Kendall, K.C., and K.S. McKelvey. (2008) Hair collection. pp141-182, In *Noninvasive Survey Methods for Carnivores* (R.A. Long, P. MacKay, W.J. Zielinski, and J.C. Ray, eds), Island Press, Washington DC.

Long, R.A., MacKay, P., Zielinski, W.J. and J.C. Ray. (2008) *Noninvasive Survey Methods for Carnivores*, Island Press, Washington, DC.

Mowat, G., and D. Paetkau (2002) Estimating marten *Martes americana* population size using hair capture and genetic tagging. *Wildlife Biology*, 8:201-209.

Pocock, M.J.O, and N. Jennings. (2006) Use of hair tubes to survey for shrews: new methods for identification and quantificationof abundance. *Mammal Review*, 36:299-308.

QIAamp® DNA Mini and Blood Mini Handbook, 3rd edition. Available at: http://www.qiagen.com/literature/render.aspx?id=200373

Vaughan, T., Ryan, J.M., and N. Czaplewski. (2010) *Mammalogy,* 5th edition, Jones and Bartlett publishers, Sudbury, MA.

Instructors Resources

Equipment and other resources for completing the exercises in this manual can be purchased from a variety of vendors, including those listed below. Note: The inclusion of any manufacturer on this list should in no way be interpreted as an endorsement of their products.

GENERAL FIELD EQUIPMENT SOURCES

FORESTRY SUPPLIERS INC.

205 West Rankin Street
P.O. Box 8397
Jackson, MS 39284-8397 USA
Sales: 800-647-5368
International: 601-354-3565

BEN MEADOWS COMPANY

PO Box 5277
Janesville, WI 53547-5277 USA
Phone: 800-241-6401
(Worldwide): 608-743-8001

ALANA ECOLOGY LTD.

New Street
Bishop's Castle
Shropshire SY9 5DQ UK
Tel: +44 (0)1588 630173
Fax: +44 (0)1588 630176

MAMMALIAN SKULLS AND SKELETONS

SKULLS UNLIMITED INTERNATIONAL

10313 S. Sunnylane Road
Oklahoma City, Ok 73160 USA
1-800-659-SKULL
www.SkullsUnlimited.com
Sales@SkullsUnlimited.com

INTERNET RESOURCES

BIOTELEMETRY RESOURCES
http://www.biotelem.org/

BIOTELEM LISTSERVER HOME PAGE
http://www.bgu.ac.il/life/bouskila/telemetry.html

WILDLIFE TELEMETRY CLEARINGHOUSE
http://www.uni-sb.de/philfak/fb6/fr66/tpw/telem telem.htm

ILLINOIS NATURAL HISTORY SURVEY WILDLIFE

ECOLOGY SOFTWARE SERVER
(telemetry data analysis programs) http://nhsbig.inhs.uiuc.edu/

CAMERA TRAPS

TRAILCAMPRO.COM

3620 South National
Springfield, MO 65807 USA
1-800-791-0660

CAMERA TRAPS CC

PO Box 2006
Hillcrest
KwaZulu-Natal
3650
chris@cameratrap.co.za
+27 83 560 0555
+27 86 505 5536
http://www.cameratrap.co.za/

BUSHNELL CORPORATION

9200 Cody
Overland Park, KS 66214-1734
Consumers - (800) 423-3537
Fax - (913) 752-3550

NON TYPICAL, INC.

PO Box 10447

Green Bay, WI 54307-0447 USA
Phone: (920) 347-3810
Fax: (920) 347-3820
http://www.cuddeback.com/

TELEMETRY EQUIPMENT MANUFACTURERS

AVM INSTRUMENT COMPANY, LTD.

2356 Research Drive, Livermore,
CA , USA 94550
Phone: +1.510.449.2286
Fax: +1.510.449.3980
email: avmtelem@ix.netcom.com

ADVANCED TELEMETRY SYSTEMS INC.

470 1st Ave. No., Box 398, Isanti,
Minnesota,USA 55040
Phone: +1.612.444.9267
e-mail: 70743.512@compuserve.com

BIOTRACK

Stoborough Croft, Grange Road,
Wareham, Dorset , UK BH20 5AJ
Phone: +44.929.552.992
Fax: +44.929.554.948

CUSTOM ELECTRONICS OF URBANA, INC.

2009 Silver CT. W.,
Urbana, Illinois, USA 61801
Phone: +1.217.344.3460
Fax: +1.217.344.3460
Custom Telemetry Co.
1050 Industrial Drive,
Watkinsville, Georgia,
USA 30677
Phone: +1.706.769.4024
Fax: +1.706.769.4026

HOLOHIL SYSTEMS LTD.

112 John Cavanagh Road, Carp,
Ontario K0A 1L0
Phone:+1.613.839.0676
Fax: +1.613.839.0675

LOTEK ENGINEERING INC.

115 Pony Drive,
Newmarket,
Ontario L3Y 7B5
Phone:+1.905.836.6680
Fax: +1.905.836.6455
e-mail: telemetry@lotek.com

MARINER RADAR LTD.

Suffolk, England NR32 5DN
Phone: +44.502.567195
Fax: +44.502.567195

MERLIN SYSTEMS, INC.

445W Ustick Rd,
Meridian, Idaho 83642
Phone:+1.208.884.3308
Fax:+1.208.888.9528
E-Mail: merlin@cyberhighway.net

MICROWAVE TELEMETRY INC.

10280 Old Columbia Road,
Suite 260,
Columbia, Maryland,
USA 21046
Phone: +1.410.290.8672
Fax: +1.410.290.8847
e-mail: Microwt@aol.com

MINI-MITTER CO., INC.

P.O. Box 3386,
Sunriver, Oregon, USA 97707
Phone: +1.503.593.8639
Fax: +1.503.593.5604
e-mail: rrushmmtr@aol.com

TELEVILT INTERNATIONAL AB

Box 53 S-711 22
Lindesberg, Sweden
Phone: +46.581.17195
Fax: +46.581.17196

TELONICS

932 East Impala Avenue,
Mesa, Arizona USA 85204-66990
Phone: +1.602.892.4444
Fax: +1.602.892.9139

WILDLIFE MATERIALS INC.

Route 1, Box 427A,
Carbondale, Illinois, USA 62901
Phone:+1.618.549.6330
Fax: +1.618.457.3340

SOUND RECORDING

THE WILDLIFE SOUND RECORDING SOCIETY

http://www.wildlife-sound.org/index.html
(many useful ideas and excellent advice)

REVIEWS OF SOUND ANALYSIS SOFTWARE

http://zeeman.ehc.edu/envs/Hopp/sound.html

RAVEN LITE SOFTWARE

http://www.birds.cornell.edu/brp/raven/RavenOverview.html

AUDACITY SOFTWARE

http://audacity.sourceforge.net/

MACAULEY LIBRARY, CORNELL LAB OF ORNITHOLOGY

http://macaulaylibrary.org/index.do?gclid=CLe929yMt6UCFYHc4Aodk0PWYg

EQUIPMENT SUPPLIERS

SAUL MINEROFF ELECTRONICS, INC.

574 Meachem Avenue,
Elmont, NY 11003 USA
Phone: (516) 775-1370
FAX: (516) 775-1371
info@mineroff-nature.com
http://www.mineroff.com/

FULL COMPASS SYSTEMS

8001 Terrace Ave.
Middleton, WI 53562

(800) 356-5844
www.fullcompass.com

MARICE STITH RECORDING SERVICES

732 Bowling Green Road
Cortland, New York 13045
(607) 756-0145
www.stithrecording.com

PROFESSIONAL SOUND SERVICES

311 W 43rd St, Suite 1100
New York NY 10036
(800) 883-1033
www.pro-sound.com

ZZOUNDS MUSIC

PO Box 479
Franklin Lakes, NJ 07417-0479
http://www.zzounds.com

ULTRASOUND RECORDING EQUIPMENT (BAT DETECTORS)

ULTRA SOUND ADVICE

27 Merton Hall Road
Wimbledon, London
SW19 3PR
United Kingdom
email sales@ultrasoundadvice.co.uk
+44 [0] 20 8287 4614
http://www.ultrasoundadvice.co.uk/

WILDLIFE ACOUSTICS

970 Sudbury Road
Concord, MA 01742-4939
(888) 733-0200 toll free
+1 (978) 369-5225 international
+1 (781) 207-5523 fax
email:support2010@wildlifeacoustics.com
http://www.wildlifeacoustics.com

PETTERSSON

Uppsala Science Park
Dag Hammarskjolds v. 34A
S-751 83 UPPSALA
Sweden
+46 1830 3880
E-mail: info@batsound.com

Web:http://www.batsound.com

BATBOX LTD

2A Chanctonfold
Horsham Road
Steyning
West Sussex
BN44 3AA
Phone: 01903 816298
emailinfo@batbox.com

TITLEY SCIENTIFIC

Email:info@titley-scientific.com
http://www.titley.com.au

About the Author

Jim Ryan, is a mammalogist and a coauthor on *Mammalogy*, a college textbook (5th edition) published by Jones & Bartlett Publishers in 2010. He is also the author of *Adirondack Wildlife, A Field Guide*, published by the University Press of New England. In addition, Professor Ryan has published numerous scientific articles in professional journals. He has conducted field research around the world, including sites in Madagascar, Kenya, Uganda, Ghana, Trinidad, and Ecuador. His research has been funded by the National Science Foundation, The National Geographic Society, and other funding agencies.

The author is currently Professor of Biology at Hobart and William Smith Colleges in Geneva, NY., where he holds the Phillip J. Moorad '28 and Margaret N. Moorad Professor of Sciences endowed chair. He received his Ph. D in zoology from The University of Massachusetts, a master's degree in biological sciences from The University of Michigan and a bachelor's degree in zoology from The State University of New York at Oswego.

Additional information about the author and about mammals can be found at his Blog at http://www.wildmammal.com.

Lightning Source UK Ltd.
Milton Keynes UK
UKOW01f2120120614

233310UK00002B/4/P